The application and numerical solution of integral equations

Monographs and textbooks on mechanics of solids and fluids
editor-in-chief: G. Æ. Oravas

Mechanics: Analysis
editor: V. J. Mizel

1. M. A. Krasnoselskii *et al.*
   Integral operators in spaces of summable functions

2. V. V. Ivanov
   The theory of approximate methods and their application to the numerical solution of singular integral equations

3. A. Kufner *et al.*
   Function spaces

4. S. G. Mikhlin
   Approximation on a rectangular grid

5. D. G. B. Edelen
   Isovector methods for equations of balance

6. R. S. Anderssen, F. R. de Hoog, M. A. Lukas
   The application and numerical solution of integral equations

The application and numerical solution of integral equations

edited by

Robert S. Anderssen
Computing Research Group/Pure Mathematics, SGS, Australian National University, Canberra, ACT 2600, and Division of Mathematics and Statistics, CSIRO, Canberra City, ACT 2601

Frank R. de Hoog
Division of Mathematics and Statistics, CSIRO, Canberra City, ACT 2601

Mark A. Lukas
Computing Research Group/Pure Mathematics, SGS, Australian National University, Canberra, ACT 2600

SIJTHOFF & NOORDHOFF 1980
Alphen aan den Rijn, The Netherlands
Germantown, Maryland, USA

Softcover reprint of the hardcover 1st edition 1980

ISBN-13: 978-94-009-9132-3 e-ISBN-13: 978-94-009-9130-9
DOI: 10.1007/978-94-009-9130-9

TABLE OF CONTENTS

FOREWORD

This publication reports the proceedings of a one-day seminar on *The Application and Numerical Solution of Integral Equations* held at the Australian National University on Wednesday, November 29, 1978. It was organized by the Computing Research Group, Australian National University and the Division of Mathematics and Statistics, CSIRO. Due to unforeseen circumstances, Dr M.L. Dow was unable to participate. At short notice, Professor D. Elliott reviewed Cauchy singular integral equations, but a paper on same is not included in these proceedings. The interested reader is referred to the recent translation of V.V. Ivanov, *The Theory of Approximate Methods and their Application to the Numerical Solution of Singular Integral Equations*, Noordhoff International Publishers, Leyden, 1976.

An attempt was made to structure the program to the extent that the emphasis was on the numerical solution of integral equations for which known applications exist along with explanations of how and why integral equation formalisms arise. In addition, the programme reflected the broad classification of most integral equations as either singular or non-singular, as either Fredholm or Volterra and as either first or second kind.

The organizers would like to take this opportunity to record their appreciation and thanks for the contributions of the speakers, the able assistance of the session chairmen (Dr Mike Osborne of the Australian National University, Canberra, Dr Alex McNabb, DSIR, Wellington, New Zealand, Professor Peter Linz, University of California, Davis, California, and Dr Joe Gani, Division of Mathematics and Statistics, CSIRO, Canberra), and the active involvement of the participants. Finally, thanks must go to both the Computing Research Group, Australian National University, and the Division of Mathematics and Statistics, CSIRO, for their financial assistance which helped cover the expenses associated with invited speakers.

The excellent work of Clementine Krayshek in tracing the figures and Anna Zalucki in typing the papers is gratefully acknowledged with sincere thanks.

R.S. Anderssen, F.R. de Hoog and M.A. Lukas
Canberra, A.C.T. 2600
November 13, 1979.

INTRODUCTION

Linear integral equations in one dimension take the form

$$\lambda u(t) + \int k(t,\tau)u(\tau)d\tau = s(t) \ , \qquad (*)$$

and are called *first kind* if $\lambda = 0$ and *second kind* otherwise. The simplicity of this equation is deceptive because it includes problems as diverse as the backward heat equation, integral transforms and Green's function methods for the analysis and solution of differential equations. Generalizations involving higher dimensions and non-linearities not only increase their applicability but also their complexity.

The main thrust of the seminar was the numerical solution of integral equations. Althoug mathematical properties of such equations are important, they were stressed only when they related directly to numerical schemes and their analysis. However, because no numerical schem is universally applicable, it is necessary to examine applications of integral equations, in order to isolate subclasses of problems on which it is worthwhile expending the time and effort to obtain reliable algorithms. Thus, a major part of these proceedings is involved in examining applications of integral equations.

Historically, a natural and effective classification for integral equations is as follows If the interval of integration in (*) is finite and the kernel $k(t,\tau)$ is integrable, then equation (*) is called *Fredholm*. When $k(t,\tau) = 0$, $\tau > t$, this special subclass is called *Volterra*. Another important class of equations arise when $k(t,\tau) = g(t,\tau)/(t-\tau)$, where $g(t,\tau)$ is integrable. In such circumstances, the integral in (*) has to be interpreted as a Cauchy principal value. Such equations are called *Cauchy singular equations* while equations with their interval of integration either semi-infinite or infinite are often referred to as *singular*.

Additional levels of classification depend on the form of the kernel $k(t,\tau)$ as mathematical and numerical properties of (*) vary greatly as $k(t,\tau)$ changes. For example, the above equations are said to be *convolution* when

$$k(t,\tau) = k(t-\tau) \ ,$$

and *weakly singular* when $k(t,\tau)$ contains an integrable singularity.

Although the above classification is very useful, it is inadequate from a numerical point of view. Numerically, it is essential to distinguish between situations when $k(t,\tau)$ is only piecewise continuous, is highly oscillatory, has an algebraic singularity, is very smooth etc., as well as between special structural properties of the kernel such as convolution, degeneracy and linearity. However, no general basis of classification for integral equations on purely numerical grounds exists at present.

Further details about the above terminology and the mathematical properties of the corresponding types of integral equations can be found in the review papers in these proceedings.

The seminar itself was structured along the general lines of the mathematical classification given above. It aimed to attract an audience which had an interest (but not necessarily extensive expertise) in the numerical solution of integral equations. This is reflected in these proceedings which attempts to motivate and review the subject as well as bootstrap the novice into the general subject area. Nevertheless, the material contained in these proceedings is such as to be of interest to the expert as well as the practitioner.

R.S. Anderssen, F.R. de Hoog and M.A. Lukas
Canberra, A.C.T. 2600
November 13, 1979.

INTEGRAL EQUATIONS — NINETY YEARS ON

David Elliott
Mathematics Department, University of Tasmania, Hobart, Tasmania

1. Introduction

We are all familiar with the phenomenon, exhibited by a nation's economy, where it goes through alternating periods of great expansion and activity (periods of "boom") followed by periods of depression and reduced activity (periods of "bust"). It is perhaps not so widely recognised, even among mathematicians, that topics in mathematics can exhibit a similar behaviour pattern. One example of this is the field of Approximation Theory. The subject "boomed" during a period of forty years or so centred on the year 1900, with the work of Chebyshev, Weierstrass, Bernstein, de la Vallée Poussin, etc. This was followed by a period of decreasing activity, which lasted about thirty years, until the advent of the computer provided a great stimulus to a renewed interest in the subject. Today, Approximation Theory is alive and well and flourishing. The thesis that I would like to develop in this lecture is that the subject of Integral Equations, having experienced a boom in the first two or three decades of this century, is again entering a "boom" period and that one of the principal reasons for this renewed interest is, again, the computer.

Before outlining the history of the development of integral equations, let me give some evidence for the claim that we are entering a "boom" period. This evidence manifests itself in various ways. One indicator is the existence of this One Day Conference, although we are lagging behind the U.K. and the U.S.A. In 1971, there was a SIAM Symposium (with the same title as this one) held in Madison, Wisconsin. The U.K. has had both a Summer School on the "Numerical Solution of Integral Equations" held in 1973 (see [13] and a Winter School on "Integral Equation Methods" held in January this year. Another indicator of interest is the number of books recently published in that subject. The past fifteen years have seen the publication of many books on various aspects of integral equations and we might note those by Baker [7], Chambers [11], Cochran [12], Gakhov [17], Green [19], Hochstadt [22],

Hoheisel [23], Moiseiwitch [32], Pogorzelski [38], Widom [48], Zabreyko et al [49]. In particular, the recently published books by Chambers [11] and Moiseiwitch [32] are aimed specifically at undergraduates in U.K. Universities, a sure indication of the increasing importance of the subject. Finally, let me remark that this year sees the publicaton of a new journal, *Integral Equations and Operator Theory* which, according to the publisher's blurb, will be *devoted to the publication of current research in Integral Equations, Operator Theory and related topics with emphasis on the linear aspects of the theory.* The publication of this journal comes ten years after volume 1 of the *Journal of Approximation Theory*, and this perhaps is an indicator of the time lag in the behaviour pattern of the two subjects.

In the next three sections I propose to give an outline of the history of the development of integral equations. Of necessity this outline is skimpy and in §3 I shall digress to show how integral equations arise from boundary value problems of potential theory.

2. A Brief History of Integral Equations (the early 19th century)

I have not yet defined an integral equation, so let us repair that omission immediately. An *Integral Equation* is an equation in which an unknown function appears behind an integral sign, and we wish to find that function. Some examples of such equations are indeed very well known.

(i) From Laplace transform theory we have that if

$$g(s) = \int_0^\infty \exp(-st)f(t)dt \quad \text{then} \quad f(t) = (1/2\pi i)\int_{a-i\infty}^{a+i\infty} \exp(st)g(s)ds ,$$

for some suitably chosen real a.

(ii) From Fourier transform theory, if

$$g(s) = \int_{-\infty}^{\infty} \exp(-ist)f(t)dt \quad \text{then} \quad f(t) = (1/2\pi)\int_{-\infty}^{\infty} \exp(ist)g(s)ds .$$

These two examples come under the heading of *integral transformations*, a topic which is widely known and extensively taught, and provides a useful technique for solving differential equations. We shall say no more about them in this lecture.

According to Kline [26], *the first conscious direct use and solution of an integral equation go back to Abel.* Abel published two papers, one in 1823 and the second in 1826 (see [1]), in which he considered a generalization of the so-called isochronous pendulum problem. Referring to Figure 1, we wish to find that curve, in the vertical plane, along which a smooth particle of mass m must be constrained to fall so that its time of descent to the lowest point 0 of the curve is a prescribed function of the vertical distance fallen. If the intrinsic equation of the curve measured from the origin 0 is given by $s = s(y)$ with $s(0) = 0$ then, from the conservation of energy, when the particle is at the point P, having started from rest at the point A which is at a height h above 0, we have

2.1 $$\frac{1}{2}\, m\left(\frac{ds}{dt}\right)^2 + mgy = \text{constant} = mgh \; .$$

If t(h) denotes the time to fall to 0 from A , then we find (see Appendix 1 and Figure 1 in that Appendix)

2.2 $$t(h) = \frac{1}{(2g)^{\frac{1}{2}}} \int_0^h \frac{s'(y)dy}{(h - y)^{\frac{1}{2}}} \; .$$

This is an integral equation for s' which Abel solved to give

2.3 $$s(y) = \frac{(2g)^{\frac{1}{2}}}{\pi} \int_0^y \frac{t(\xi)d}{(y-\xi)^{\frac{1}{2}}} \; .$$

In the case of the isochronous pendulum, t is specified to be a constant T say, and from (2.3),

2.4 $$s(y) = (2T/\pi)(2gy)^{\frac{1}{2}}$$

which is the equation of a cycloid, a fact that was well known long before Abel. In his papers, Abel solved a more general integral equation than (2.2). He showed that the solution of the integral equation

2.5 $$\int_a^x \frac{\phi(y)dy}{(x - y)^{\alpha}} = f(x) \; , \quad 0 < \alpha < 1 \; ,$$

is given by

2.6 $$\phi(y) = \frac{\sin(\pi\alpha)}{\pi} \cdot \frac{d}{dy} \left\{ \int_a^y \frac{f(x)dx}{(y-x)^{1-\alpha}} \right\} .$$

Abel's achievement in solving (2.5) is quite remarkable. In present parlance, (2.5) is an integral equation of the *first kind* whose *kernel* $(x - y)^{-\alpha}$ is *weakly singular*. Lonseth [30], in his excellent review paper, puts it this way :

> *Luckily, the equation's discoverer did not know that these were deterrents. A present day graduate student of mathematics would no doubt have learned the standard Fredholm alternative theorem* (see §4) *about linear equations in Banach spaces, which would not apply, and would probably bypass* (2.5) *as an equation which does not fit into the received wisdom. But Abel - back in the spring-time of analysis - did not know enough to be daunted.*

Integral equations of *Abel-type*, which are generalizations of (2.5), occur frequently in applications, one of the more interesting of which is perhaps seismology. They have now been extensively studied and Anderssen [2] has recently given a most thorough review of them.

After Abel, the next significant step in the development of integral equations was taken by Liouville [29], who, in 1837, considered the initial value problem

2.7 $$\phi''(x) + [\rho^2 - \sigma(x)]\phi(x) = 0 \text{ with } \phi(a) = 1,\ \phi'(a) = 0$$

and showed, by using the method of variation of parameters on $\phi''(x) + \rho^2\phi(x) = \sigma(x)\phi(x)$ assuming the right hand side to be known, that it could be transformed into what we now call a *Volterra integral equation of the second kind*. These equations are of the form

2.8 $$\phi(x) - \int_a^x K(x,y)\phi(y)dy = f(x) \ ,$$

and from the differential equation (2.7) we find that $K(x,y) = (1/\rho)\sigma(y)\sin\rho(x-y)$ and $f(x) = \cos\rho(x-a)$. Liouville solved his integral equation by the method of successive approximation thereby anticipating, by many years, the methods used by Neumann [34] and Volterra [46] (see §4).

Perhaps the physical problems which have had the greatest influence on the development of integral equations are those which give rise to potential theory. Such problems arise from various sources, the mathematical theories of hydrodynamics and elasticity being just two of them. At this point I would like to digress from the historical development (to which we shall return in §4) and consider how integral equations arise in the particular case of two-dimensional potential problems.

3. Integral Equations from Two Dimensional Potential Theory

Essentially we are concerned with finding solutions of Laplace's equation $\nabla^2 u = 0$, in two dimensions, with various boundary conditions. First we consider the *interior Dirichlet problem*. Suppose the plane is divided into a bounded region D^+ and an unbounded region D^- by a simple, smooth, closed curve C. Then we want to find the real function u such that

3.1 $$\begin{cases} \nabla^2 u = 0 \ , \text{ at all points of } D^+ \ , \\ u = F \ , \text{ on } C \ , \end{cases}$$

where F is a given real function. We shall consider three integral equation formulations of this problem.

The method of the *simple layer potential* depends upon the observation that the function $\log r$, where

$$r = \{(x - x_0)^2 + (y - y_0)^2\}^{\frac{1}{2}} = |z - \tau| \text{ where } z = x_0 + iy_0 \ , \ \tau = x + iy \ ,$$

satisfies Laplace's equation. This suggests that we look for a solution of (3.1) by writing

3.2 $$u(z) = \int_C \log |z - \tau|\mu(\tau)d\tau$$

for any $z \in D^+$, where μ is a function to be determined. Now, as z approaches a point t of C through points of D^+ we obtain an integral equation of the first kind for μ given by

$$\int_C \log|t - \tau| \, \mu(\tau)d\tau = F(t) \ , \quad t \in C \ . \tag{3.3}$$

Referring to Figure 2, let P_0 denote the point t, let P be an arbitrary point τ on C, and write $d\tau = \exp(i\theta)ds$, where s denotes the arc length measured along C from some fixed point, so that $0 \leqslant s \leqslant l$. If we write $\mu(\tau) = \phi(s)$, $F(t) = f(s_0)$ and $K(s_0,s) = \exp(i\theta)\log r$, then (3.3) becomes

$$\int_0^l K(s_0,s)\phi(s)ds = f(s_0) \ , \quad 0 \leqslant s_0 \leqslant l \ . \tag{3.4}$$

Although K is discontinuous when $s = s_0$, nevertheless it is square integrable. An equation such as (3.4) is known as a *Fredholm integral equation of the first kind*. Green (who considered the Dirichlet problem before Dirichlet did), in his 1818 essay (see [20, p.10] when considering the 3 dimensional analogue of (3.4) commented :-

> *But we have no general theory for equations of this description and whenever we are enabled to resolve one of them, it is because some consideration peculiar to the problem renders, in that particular case, the solution comparatively simple, and must be looked upon as the effect of chance, rather than of any regular and scientific procedure.*

This was published five years after Abel's first paper and it is interesting to speculate as to whether Green knew of Abel's paper and if so whether he is ascribing Abel's solution to "chance rather than of any regular and scientific procedure"!

Both our second and third approaches to the solution to (3.1) are based on the use of the so-called *Cauchy integral*. If we define

$$w(z) = \frac{1}{\pi i}\int_C \frac{\mu(\tau)d\tau}{\tau - z} \tag{3.5}$$

where we assume μ to be *real*, then this function satisfies Laplace's equation at all points of D^+ and, furthermore, as z approaches a point t of C through the region D^+ then, by the Sokhotski-Plemelj formula (see, for example, [17]), we have that

$$w(t) = \mu(t) + \frac{1}{\pi i} ⨍_C \frac{\mu(\tau)d\tau}{\tau - t} \ , \quad t \in C \ , \tag{3.6}$$

where $⨍_C$ denotes a Cauchy principal value integral taken over C. We now distinguish two cases, depending upon whether we look upon u as the real or imaginary part of w.

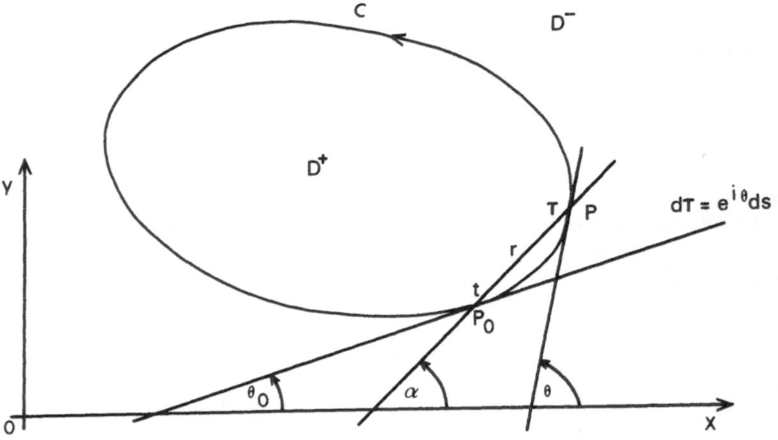

FIGURE 2.

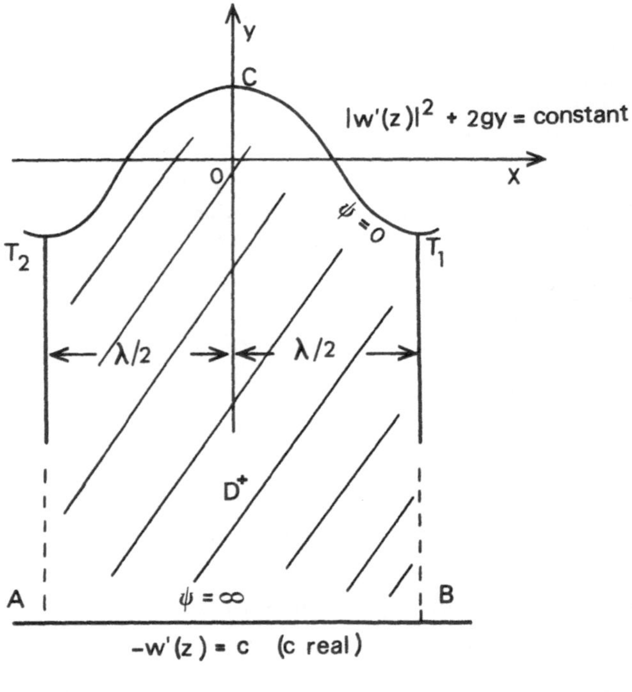

FIGURE 3.

u = Re w

Suppose first that $u = Re\ w$. On taking real parts in (3.6), recalling that we assumed both μ and F to be real, and that $u = F$ on C, we find

3.7
$$F(t) = \mu(t) + \frac{1}{\pi} \int_C \mu(\tau)\ Im(\frac{d\tau}{\tau - t}) \ .$$

Since (see Figure 2), $d\tau = e^{i\theta}ds$ and $\tau - t = r\ e^{i\alpha}$, (3.7) becomes

3.8
$$\mu(t) + \frac{1}{\pi} \int_0^l \frac{\sin(\theta-\alpha)}{r} \mu(\tau)ds = F(t) \ .$$

Again, on putting $\mu(t) = \phi(s_0)$, $\mu(\tau) = \phi(s)$, $F(t) = f(s_0)$ and defining $K(s_0,s) = \sin(\theta-\alpha)/r$, (3.8) becomes

3.9
$$\phi(s_0) + \frac{1}{\pi} \int_0^l K(s_0,s)\phi(s)ds = f(s_0) \ ,$$

$0 \leqslant s_0 \leqslant l$. It remains to investigate the nature of the kernel K. For the curve C as defined above, K is a continuous function of both s_0 and s except perhaps when $s = s_0$ where the above definition is indeterminate. If we define $K(s_0,s_0) = \lim_{s\to s_0} K(s_0,s)$ then we find (see Appendix 2) that $K(s_0,s_0) = \kappa_0/2$, where κ_0 is the curvature of C at the point P_0. Thus, provided C has finite curvature at each point, K will be continuous and the Cauchy principal value integral can be replaced by an ordinary (Riemann) integral. Any equation of the form

3.10
$$\phi(x) - \lambda \int_a^b K(x,y)\phi(y)dy = f(x) \quad ,$$

$a \leqslant x \leqslant b$, is known as a (non-singular) *Fredholm integral equation of the second kind.* This approach is known as that of the *double layer potential.* Using the terminology of hydrodynamics in the first method we replaced the curve C by a distribution of *sources,* in the second method by a distribution of *doublets.*

There are some interesting observations to be made on these examples. Firstly, we observe that what was originally a two-dimensional problem in u has been replaced by a one-dimensional one for the unknown function ϕ. Once ϕ has been found we can find u for any point of D^+ using (3.2) in the simple layer formulation, or $u = Re\ w$ from (3.5) in the double layer formulation. Since ϕ is a function of a single variable only, the complexity of our original problem is diminished. (We might note that if we attempt a similar analysis for a three-dimensional problem, we reduce it to a two-dimensional integral equation). We can also see that integral equations will arise in problems in which we require the effect (at a distance say) due to a certain cause. On the assumption that we can superpose the effects due to several causes, we consider a certain distribution over C of "simpler" causes and then sum, by means of integration, to obtain the required effect. It is a technique that has proved very effective in many contexts.

$u = Im\ w$

Returning again to (3.6), let us suppose that $u = \mathrm{Im}\, w$, then on taking imaginary parts of each side of (3.6) we find, since μ is real, that

3.11 $$F(t) = -\frac{1}{\pi}\int_C \mu(\tau)\mathrm{Re}\left(\frac{d\tau}{\tau - t}\right) = -\frac{1}{\pi}\int_0^l \frac{\cos(\theta-\alpha)}{r}\mu(\tau(s))ds \ .$$

If we now choose θ (see Figure 2) to be the new variable of integration, so that $d\theta/ds = \kappa$, where κ is the curvature, then on writing $F(t) = -f(\theta_0)$, $\mu(\tau(s)) = \phi(\theta)$, (3.11) becomes

3.12 $$\frac{1}{\pi}\int_0^{2\pi}\left(\frac{\cos(\theta-\alpha)}{\kappa\ r}\right)\phi(\theta)d\theta = f(\theta_0)\ , \quad 0 \leqslant \theta_0 \leqslant 2\pi\ .$$

It is obvious that the kernel of this equation will be infinite when $\theta = \theta_0$. However, if we write

3.13 $$\frac{\cos(\theta-\alpha)}{\kappa\ r} = \frac{1}{2}\cot((\theta-\theta_0)/2) + k(\theta,\theta_0)$$

then it can be shown, under certain conditions (see Appendix 2), that k is continuous. (In the particular case when C is a circle, $k \equiv 0$). Equation (3.12) can be rewritten as

3.14 $$\frac{1}{2\pi}\int_0^{2\pi}\cot\left(\frac{\theta-\theta_0}{2}\right)\phi(\theta)d\theta + \frac{1}{\pi}\int_0^{2\pi}k(\theta,\theta_0)\phi(\theta)d\theta = f(\theta_0)\ ,$$

$0 \leqslant \theta_0 \leqslant 2\pi$, where the first integral must be retained as a Cauchy principal value integral. Equation (3.14) is known as *a Hilbert equation of the first kind*, and comes into that class of integral equations loosely described as *singular*.

So far we have considered only linear integral equations, but we shall now see how a potential problem, with a non-linear boundary condition, can give rise to a non-linear integral equation. The problem is that of finding the shape of a symmetric wave of constant form travelling along the surface of an incompressible fluid, of infinite depth, with constant velocity. If we assume that the flow is two-dimensional and that the motion is irrotational then, on reducing the wave to rest, we wish to find a complex potential $w = \phi + i\psi$ which is an analytic function in the interior of the region ABT_1CT_2A and satisfies the boundary conditions as shown in Figure 3. We assume that the wave has wavelength λ and that it travels with speed c. We require the shape of the curve T_1CT_2 which is assumed to be symmetric about Oy. The details of the analysis are to be found in Milne-Thomson [31, Chapter XIV], where it is shown that as an intermediate step of the analysis we need to solve the non-linear integral equation

3.15 $$\Phi(t) = \frac{\mu}{3}\int_0^{\pi}\frac{1}{\pi}\log\left|\frac{\sin(t+\tau)/2}{\sin(t-\tau)/2}\right| \frac{\sin\Phi(\tau)}{\{1+\mu\int_0^{\tau}\sin\Phi(\xi)d\xi\}}d\tau\ ,$$

μ being a constant. This equation, known in the literature as *Nekrasov's equation*, is frighteningly non-linear, but appears to yield well to numerical techniques for an approximate solution. However, let us observe that in this problem, which concerns an unbounded two-dimensional region with an unknown boundary, the integral equation formulation reduces the problem to that of finding a function of a single variable defined on the finite interval $[0, \pi]$. This simplification is not inconsiderable.

We have seen, in this section, how the potential problem gives rise to various integral equations with which we have associated the names of Volterra, Fredholm and Hilbert. Let us now look at the contributions these men made to integral equations during the first "boom" period which started around 1895 and subsided about 30 years later.

4. THE WORK OF VOLTERRA, FREDHOLM, HILBERT AND OTHERS

Volterra apparently first met with an integral equation in 1884 when dealing with the problem of distribution of electric charges on a segment of the surface of a sphere. However, it was not until 1896 that he seriously studied them (see [46]) and he considered equations of both the first and second kind, given respectively by

4.1 $$\int_a^x K(x,y)\phi(y)dy = f(x) ,$$

and $$\phi(x) - \int_a^x K(x,y)\phi(y)dy = f(x) .$$

He realised that these equations could be looked upon as the limiting case, as $n \to \infty$, of n linear algebraic equations in n unknowns, the equations being of *triangular form*. This suggested to him that the solution of the equation of the second kind could be written as

4.2 $$\phi(x) = f(x) + \int_a^x R(x,y)f(y)dy ,$$

and he found that the *resolvent kernel R* is given by

4.3 $$R(x,y) = \sum_{m=0}^{\infty} K_{m+1}(x,y) ,$$

where the *iterated kernels* K_m are defined recursively by

4.4 $$\begin{cases} K_1(x,y) = K(x,y) , \\ K_{m+1}(x,y) = \int_a^x K(x,z)K_m(z,y)dz, \quad m = 1,2,3,\ldots . \end{cases}$$

The next significant development is due to Fredholm [16, pp.81-106] who considered the equation

4.5 $$\phi(x) - \lambda \int_a^b K(x,y)\phi(y)dy = f(x) \ , \quad a \leqslant x \leqslant b \ ,$$

where he assumed that K was bounded and integrable. On comparing this equation with the second of (4.1), we observe that the upper limit of integration x has been replaced by b, and furthermore the scalar parameter λ has been introduced. These seemingly minor changes have a profound effect on the subsequent analysis. Prior to Fredholm, Neumann [34] in 1877 found a solution of (4.5) in the form

4.6 $$\phi(x) = f(x) + \lambda \int_a^b R(x,y)f(y)dy$$

where $R(x,y) = \sum_{m=0}^{\infty} \lambda^m K_{m+1}(x,y)$, the iterated kernels being defined as in (4.4) with b replacing x. However, Neumann's series converged only for sufficiently small values of $|\lambda|$. It was Poincaré who suggested that the resolvent function R may be a *meromorphic function* of λ. If we recall Cramer's rule and apply it to the solution of n linear algebraic equations of the form $(I-\lambda A)\underline{x} = \underline{f}$, then $\underline{x}$ will be given as the ratio of two *polynomials* of degree $\leqslant n$ in λ. Fredholm put these ideas together and, by considering the limiting behaviour as $n \to \infty$ of n linear algebraic equations, showed that the resolvent R is given by $R(x,y;\lambda) = D(x,y;\lambda)/D(\lambda)$, provided $D(\lambda) \neq 0$. If

4.7 $$K\begin{pmatrix} x_1,x_2,\ldots,x_m \\ y_1,y_2,\ldots,y_m \end{pmatrix} = \det(K(x_i,y_j)) \ , \quad i,j = 1(1)m \ ,$$

then we have

4.8 $$\begin{cases} D(\lambda) = 1 + \sum_{m=1}^{\infty} \frac{(-\lambda)^m}{m!} \int_a^b \cdots \int_a^b K\begin{pmatrix} \xi_1,\xi_2,\ldots,\xi_m \\ \xi_1,\xi_2,\ldots,\xi_m \end{pmatrix} d\xi_1 \ldots d\xi_m \\ \text{and} \\ D(x,y;\lambda) = K(x,y) + \sum_{m=1}^{\infty} \frac{(-\lambda)^m}{m!} \int_a^b \cdots \int_a^b K\begin{pmatrix} x,\xi_1,\ldots,\xi_m \\ y,\xi_1,\ldots,\xi_m \end{pmatrix} d\xi_1 \ldots d\xi_m \ . \end{cases}$$

Each power series converges for all finite values of $|\lambda|$. In the particular case of Volterra's equation of the second kind we have $D(\lambda) = \exp(-A_1\lambda)$ where $A_1 = \int_a^b K(x,x)dx$, so that in this case $D(\lambda)$ has no zeros. The extension of Fredholm's theory to the case when the kernel is square integrable (i.e. such that $\int_a^b \int_a^b |K(x,y)|^2 dxdy$ exists) was given by Carleman [9] in 1921.

Arising from Fredholm's analysis we have the famous *Fredholm alternative theorem*, a shortened form of which is as follows: *either* (4.5) possesses a unique solution for all f, the corresponding homogeneous equation ((4.5) with $f \equiv 0$) possesses a non-trivial solution. With this theorem Fredholm was able to consider the existence of solutions of the interior Dirichlet problem. This problem has previously been reformulated as an integral equation (see (3.9)). From other arguments we know that when $f \equiv 0$ the only solution of (3.9) is given by $\phi \equiv 0$. Thus from Fredholm's alternative and his analysis, if K is continuous then a solution of (3.9) exists and is unique for arbitrary f. As we have seen, K is continuous if C has finite curvature at each point. Thus for a wide class of boundary curves C, Fredholm showed [16, pp.61-68] that *uniqueness* of the solution of the homogeneous problem *implies existence* of the solution of the interior Dirichlet problem; a remarkable result.

We must now turn to the work of Hilbert. Hermann Weyl [47] tells us that when, in the winter of 1900-1901, Hilbert heard of the work that Fredholm was doing (in Stockholm), he *caught fire*. The result of this "combustion" was a series of 6 papers written in the period from 1904 to 1910, and later gathered together in book form, see [21]. There appear to have been two main influences on Hilbert's studies into integral equations. One was the Sturm-Liouville theory of self-adjoint differential equations whose solutions can be given in terms of expansions involving the eigenfunctions of the differential operator; the other was the theory of quadratic forms of matrices. The former led to a consideration of integral equations with real, symmetric kernels ($K(x,y) = K(y,x)$, K real) while from the latter there is the theory of orthogonal transformations. Hilbert thoroughly explored the consequences of both these areas. Suppose the equation

4.9
$$\lambda_k \int_a^b K(x,y)\phi_k(y)dy = \phi_k(x)$$

possesses a non-trivial solution. Then λ_k is known as a *characteristic value* of K with ϕ_k its corresponding *eigenfunction*. (We reserve the word *eigenvalue* for $1/\lambda_k$). For real, symmetric K it can be shown that (4.9) possesses at least one characteristic value, but no more than a countable number. Furthermore, the characteristic values are all real and eigenfunctions corresponding to distinct characteristic values will be orthogonal, so that if $\lambda_j \neq \lambda_k$ then the *inner product* (ϕ_j,ϕ_k), defined by $(\phi_j,\phi_k) = \int_a^b \phi_j(x)\phi_k(x)dx$, is zero. Among other things, Hilbert showed that if λ is not a characteristic value of the real, symmetric kernel K, then the solution of (4.5) is given by

4.10
$$\phi(x) = f(x) + \lambda \sum_{k=1}^{\infty} \frac{(\phi_k,f)}{\lambda_k - \lambda} \phi_k(x) ,$$

the series converging absolutely and uniformly for all $x \in [a,b]$ when K is continuous. Although Tricomi [45] describes Fredholm's solution as "elegant", this solution is surely aesthetically more pleasing.

The papers by Fredholm and Hilbert have had a profound effect on the development of mathematics this century. From the subsequent work done by Schmidt, Reisz, Fréchet and Banach the subject we know today as *functional analysis* has developed. For an excellent survey of this development, see Bernkopf [8]. But, before we leave Hilbert, it is not without interest to note that in the second half of his life when he turned his attention to Physics, he *applied* integral equations to both the kinetic theory of gases and to the theory of radiation (see Weyl [47, p.653]).

In our brief review of the history of integral equations we have so far considered only non-singular integral equations. However, singular integral equations of the form

4.11 $$a(t)\phi(t) + \frac{b(t)}{\pi i}\int_C \frac{\phi(\tau)d\tau}{\tau - t} + \lambda\int_C K(t,\tau)\phi(\tau)d\tau = f(t) \ , \ t \in C \ ,$$

are also of some importance. The foundations of the theory of these equations were laid down, almost simultaneously but independently, by Hilbert [21] and Poincaré [39] in the first decade of this century. However, the development of the theory made less rapid progress than that of non-singular equations. It was not until 1921 that Noether [36] published his theorems which are for singular equations, what the Fredholm theory is for non-singular ones. A year later Carleman [10] gave his solution to (4.11) in the particular case when b is a constant, $K \equiv 0$ and C is the arc $(-1,1)$. For the next two or three decades, progress in the theory was due entirely to Russian mathematicians and mention must be made of the contribution of Muskhelishvili (see [33]), whose interest in these equations was aroused by his work in the theory of elasticity.

5. Integral Equations and the Role of the Numerical Analyst

For the numerical analyst, integral equations lie at the confluence of two main streams of development of his subject. One stream of development is the quadrature problem (the approximation of an integral by a finite sum); the other is the numerical solution of algebraic equations, both linear and non-linear. The past two or three decades have seen vigorous growth in both these areas, and one might postulate that if integral equations had not been with us for the past 150 years, then numerical analysts would have invented them! Of the two main streams noted above, perhaps the more important has been that of being able to solve numerically systems of algebraic equations. Applied mathematicians, physicists, engineers, etc., can now think in terms of using integral equations as mathematical models since the drudgery of finding approximation solutions has been removed by the computer. It is easy to forget how tedious computation was prior to the computer age. This misery is graphically described by Neville Shute (Norway) in his autobiography [37, Chapter 3]. In 1925 he was involved with stress calculations for the framework of the R100 airship. He used an iteration process to solve a system of algebraic equations and each calculation *...usually occupied two calculators about a week, using a Fuller slide rule and working in pairs to check for arithmetical mistakes.* The work was so tedious that when a calculation was finished, *...it produced a satisfaction almost amounting to a religious experience.* The calculation incidentally involved *the solution of a lengthy simultaneous equation containing up to seven unknown quantities*; (Shute does not say whether these equations were linear or non-linear).

But to return to the solution of integral equations; the first step is to discretize the equation in some way to produce a system of algebraic equations. There are various ways of discretization and I shall leave it to others today to describe these. But, having discretized to give a system of say n equations, we are interested in having computable bounds on the error between the approximate solution and the exact solution, and also in knowing whether the approximate solution *converges* in the limit as $n \to \infty$ to the exact solution. It is of course, the problem of Volterra, Fredholm and Hilbert all over again! The problem of convergence is mathematically fascinating and has attracted much attention. An analysis dating back to about 1940 was given by Kantorovich and Krylov (see [25]). This was followed by the theory of *collectively compact operators* which forms the subject of the book by Anselone [3]. Most recently, the use of *restriction* and *prolongation operators* has been advocated by Noble [35] and subsequently developed by some of his students (Spence [42] [43],Thomas [44] and Linz [28]).

I have digressed a little with convergence; the state of the art (up to mid 1975) of computing approximate solutions of integral equations is given in the recently published, encyclopaedic book by Baker [7]. In spite of its size, one can see at least four areas where more work remains to be done. These are

(i) the production of robust procedures for the computer solution of integral equations;

(ii) the development of methods for singular integral equations with Cauchy kernel;

(iii) development of methods for non-linear integral equations; and

(iv) extensions of algorithms and analysis for higher dimensional integral equations.

Let me comment briefly on each of these areas.

A month or so ago I consulted the International Mathematical Software Library (IMSL) for procedures for solving integral equations, and found nothing! Miller [13], in his lecture to the 1973 Summer School on the Numerical Solution of Integral Equations, found at that time only 3 procedures written in Algol. Two (Pouzet [40], Rumyantsev [41]) were for equations of Volterra type and one (Elliott & Warne [15]) was for linear Fredholm or linear Volterra equations ot the second kind. Subsequently, Atkinson [5], published in 1976 a Fortran procedure for Fredholm equations of the second kind. Miller, in a recent letter, tells me that he has now put a couple of procedures for Fredholm integral equations into the NAG library, but that seems to be about it. Users of integral equations have not been as well served computationally as those of ordinary differential equations for whom the Runge-Kutta-Gill subroutine [18] has been available since the early 1950s. If one had formulated a differential equation, then one could always attempt to get numbers, although not necessarily with great accuracy or efficiency, by using this routine as a "black box". It encouraged the use of differential equations, but there has been nothing comparable for integral equations.

Both Fredholm equations of the first kind, and singular equations with Cauchy kernel have generated an aura of mystery about them which still seems to persist in the latter case. However, in the former case the difficulties associated with the approximate solution of such equations are now well understood, although it is arguable whether they can ever be satisfactorily overcome, since the problem is ill-posed. A very recent paper by Lee and Prenter [27] contains both an extensive bibliography on this topic and an analysis for algorithms based on "filtering" techniques. But, for singular equations with Cauchy kernels much remains to be done. Baker [7] devotes only 7 pages (out of about 1,000) to the approximate solution of these equations, so it is particularly interesting to note that two papers today are devoted to them. Now that the book by Ivanov [24] is available in an English translation, this should stimulate wider interest in algorithms for these equations, and I predict that within the next five years such equations will no longer be considered in any way "difficult".

But surely the greatest thrust in the future will come from the solution of non-linear integral equations. At a seminar on such equations in 1964 [4], Noble observed, *as a rough generalization it would seem that, as physical problems become more complicated, integral equations become more useful.* Since Nature seems to thrive on non-linearity, it seems that non-linear integral equations must become of increasing importance. The three Chapters by Rall in [13] form a good introduction to the problem of finding approximate solutions to such equations. There are some interesting problems associated with non-linear equations, not the least of which is the *bifurcation problem* on which some numerical work has recently been done by Atkinson [6].

Finally, since many problems reduce to integral equations in which the unknown function is of two or more independent variables, it is essential that algorithms be developed for the efficient solution of *multi-dimensional* integral equations. The use of finite elements for such problems has a certain attraction that will surely be fully investigated in years to come.

As far as the approximate solution of integral equations is concerned, much has been done but much remains to be done. It does appear though, that the numerical analyst has a key role to play if integral equations are to be widely accepted and extensively used in the future.

6. Why Ninety Years On?

I have called this lecture, *Integral Equations - Ninety Years On,* and an immediate response, I suppose, is to ask what happened in 1888? The brief answer to that is, "Very little", but I have picked on this year because it is the one in which the phrase *integral equations* was first used. It appeared in a paper by duBois - Reymond [14], but its acceptance was far from immediate. Thus Volterra's work was on *fonctionelles,* and Fredholm's 1903 paper [16] was entitled *Sur Une Classe d'Équations Fonctionelles.* But the P.R. man for *integral equations* appears to have been Hilbert who in 1904 (see [21]) wrote the first of his six papers on *linearen Integralgleichungen,* and *Integralgleichungen* it has been ever since for German mathematicians, although Plemelj did publish a paper in the same year in which he referred to *Fredholmschen Funktionalgleichungen.* However, international acceptance of *integral equations* seemed to follow very rapidly after Hilbert's paper. Thus Goursat, in 1907, wrote on *...des équations intégrales.* A slim treatise by Bocher in 1909 was on the *...study of Integral Equations,* and a paper by Fubini in 1910 referred to *Equazioni integrale.* And that is what they have been ever since.

The past ninety years have indeed seen some very exciting developments in both the theory and application of integral equations. Whether the renewed interest which is currently being generated will develop into a genuine "boom" or not, only time will tell. The only certainty is the rather melancholy observation that none of us listening here today will ever completely know what the *next* ninety years will bring forth!

REFERENCES

[1] Abel, N.H. *Oeuvres Complètes*, 1, 11-27 and 97-101, Johnson Reprint Corporation, New York, 1973.

[2] Anderssen, R.S. Application and numerical solution of Abel-type integral equations, University of Wisconsin-Madison, Mathematics Research Center, Technical Summary Report #1787, 1977.

[3] Anselone, P.M. Collectively Compact Operator Approximation Theory, Prentice-Hall, Englewood Cliffs, N.J., 1971.

[4] Anselone, P.M. (ed.). Nonlinear Integral Equations, The University of Wisconsin Press, Madison, 1964.

[5] Atkinson, K.E. An automatic program for linear Fredholm integral equations of the second kind, *ACM Trans. on Math. Software*, 2, 154-171 and 196-199, 1976.

[6] Atkinson, K.E. The numerical solution of a bifurcation problem, *SIAM J. Numer. Anal.*, 14, 584-599, 1977.

[7] Baker, C.T.H. The Numerical Treatment of Integral Equations, Clarendon Press, Oxford, 1977.

[8] Bernkopf, M. The development of function spaces with particular reference to their origins in integral equation theory, *Arch. Hist. Exact Sci.*, 3, 1-96, 1966.

[9] Carleman, T. Zur Theorie der linearen Integralgleichungen, *Math. Z.*, 9, 196-217, 1921.

[10] Carleman, T. Sur la resolution de certaines équations intégrales, *Ark. Mat.*, 16, No. 26,(19 pages), 1922.

[11] Chambers, L.G. Integral Equations: A Short Course, Int. Textbook Co., London, 1976.

[12] Cochran, J.A. The Analysis of Linear Integral Equations, McGraw-Hill, New York, 1972.

[13] Delves, L.M. and Walsh. J. (eds.). Numerical Solution of Integral Equations, Clarendon Press, Oxford, 1974.

[14] Du Bois-Reymond, P. Bemerkungen über $\Delta z = 0$, *Jour. für Math.*, 103, 204-229, 1888.

[15] Elliott, D. and Warne, W.G. An algorithm for the numerical solution of linear integral equations, *Int. Comput. Cent. Bull.*, 6, 207-224, 1967.

[16] Fredholm, I. *Oeuvres Complètes*, Malmö, 1955.

[17] Gakhov, F.D. Boundary Value Problems, Pergamon Press, Oxford, 1966.

[18] Gill, S. A process for the step-by-step integration of differential equations in an automatic digital computing machine, *Proc. Camb. Phil. Soc.*, 47, 96-108, 1951.

[19] Green, C.D. Integral Equation Methods, Nelson, London, 1969.

[20] Green, G. An essay on the application of mathematical analysis to the theories of electricity and magnetism, see *Mathematics Papers* of the late George Green (ed. by N.M. Ferrers), Chelsea, New York, 1970.

[21] Hilbert, D. Grundzüge einer allgemeinen Theorie der linearen Integralgleichungen, 1912, reprinted by Chelsea, New York, 1953.

[22] Hochstadt, H. Integral Equations, Wiley, New York, 1973.

[23] Hoheisel, G. Integral Equations, Nelson, London, 1967.

[24] Ivanov, V.V. The Theory of Approximate Methods and Their Application to the Numerical Solution of Singular Integral Equations, Noordhoff, Leyden, 1976.

[25] Kantorovich, L.V. and Krylov, V.I. Approximate Methods of Higher Analysis, Noordhoff, Groningen, 1958.

[26] Kline, M. Mathematical Thought from Ancient to Modern Times, Oxford University Press, New York, 1972.

[27] Lee, J.W. and Prenter, P.M. An analysis of the numerical solution of Fredholm integral equations of the first kind, *Numer. Math.*, 30, 1-23, 1978.

[28] Linz, P. A general theory for the approximate solution of operator equations of the second kind, *SIAM J. Numer. Anal.*, 14, 543-554, 1977.

[29] Liouville, J. Sur le développement des fonctions, *J. Math. Pures Appl.* (1), 2, 16-35, 1837.

[30] Lonseth, A.T. Sources and applications of integral equations, *SIAM Review*, 19, 241-278, 1977.

[31] Milne-Thomson, L.M. Theoretical Hydrodynamics, Macmillan, London, 5th edition, 1968.

[32] Moiseiwitsch, B.L. Integral Equations, Longman, London, 1977.

[33] Muskhelishvili, N.I. Singular Integral Equations, Noordhoff, Groningen, 1953.

[34] Neumann, C. Untersuchungen über das logarithmische und Newtonsche Potential, Teubner, Leipzig, 1877.

[35] Noble, B. Error analysis of collocation methods for solving Fredholm integral equations, see *Topics in Numerical Analysis* (ed. J.H.H. Miller), Academic Press, London, 211-232, 1973.

[36] Noether, F. Über eine klasse singulärer integralgleichungen, *Math. Ann.*, 82, 42-63, 1921.

[37] Norway, N. (Shute). Slide Rule, Heinemann, London, 1954.

[38] Pogorzelski, W. Integral Equations and Their Applications, Pergamon Press, Oxford, 1966.

[39] Poincaré, H. *Leçons de mécanique céleste*, 3, Paris, 1910.

[40] Pouzet, P. Algorithme de résolution des équations intégrales de type Volterra par des méthodes par pas, *Chiffres*, 7, 169-173, 1964.

[41] Rumyantsev, I.A. Programme for solving a system of Volterra integral equations of the second kind, *USSR Comp. Math. and Math. Phys.*, 5, 218-224, 1965.

[42] Spence, A. On the convergence of the Nÿstrom method for the integral equation eigenvalue problem, *Numer. Math.*, 25, 57-66, 1975.

[43] Spence, A. Error bounds and estimates for eigenvalues of integral equations, *Numer. Math.*, 29, 133-147, 1978.

[44] Thomas, K.S. On the approximate solution of operator equations, *Numer. Math.*, 23, 231-239, 1975.

[45] Tricomi, F. Integral Equations, *Interscience*, New York, 1957.

[46] Volterra, V. *Opere Mathematiche*, 2, Acad. Naz. Lincei, Rome, 216-275, 1956.

[47] Weyl, H. David Hilbert and his mathematical work, *Bull. Amer. Math. Soc.*, 50, 612-654, 1944.

[48] Widom, H. Lectures on Integral Equations, Van Nostrand, New York, 1969.

[49] Zabreyko, P.P. et al., Integral Equations - A Reference Text, Noordhoff, Leyden, 1975.

Appendix 1

On referring to Figure 1, when particle is at P we have from the conservation of energy that

A1.1
$$\frac{1}{2}m\left(\frac{ds}{dt}\right)^2 + mgy = \text{constant} = mgh ,$$

so that

A1.2
$$dt = \pm \frac{ds}{(2g)^{\frac{1}{2}}(h - y)^{\frac{1}{2}}} .$$

Let $s = s(y)$, with $s(0) = 0$, denote the intrinsic equation of the path described by the particle P. If the particle starts from rest at the point A, at a height h above the origin 0, and if $t(h)$ denotes the time of descent to 0, we have from (A1.2) that

A1.3
$$t(h) = \frac{1}{(2g)^{\frac{1}{2}}} \int_0^h \frac{s'(y)dy}{(h - y)^{\frac{1}{2}}} ,$$

the negative sign being chosen in (A1.2) since $s'(y) > 0$ and t must be positive. From Abel's solution (2.6), we find that

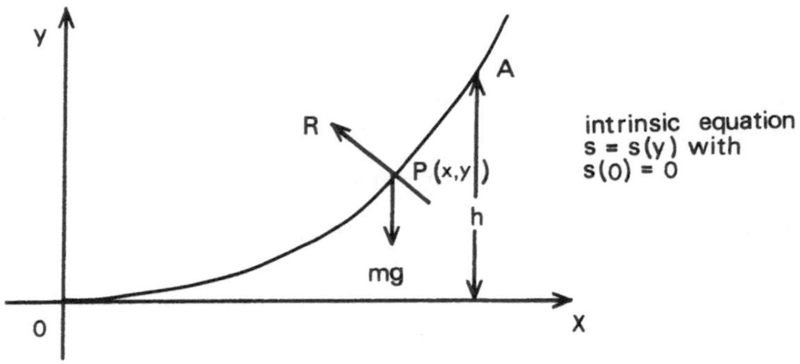

FIGURE 1.

A1.4 $$s(y) = \frac{(2g)^{\frac{1}{2}}}{\pi} \int_0^y \frac{t(h)dh}{(y-h)^{\frac{1}{2}}} ,$$

since $s(0) = 0$. If we consider the special case in which t is a constant, T say, independent of h, then we find from (A1.4) that

A1.5 $$s(y) = (2T/\pi).(2gy)^{\frac{1}{2}} .$$

Now this is the equation of an arc of a cycloid which may be obtained geometrically by considering a fixed point on the circumference of a circle of radius gT^2/π^2 as it rolls along the (underside) of the line $y = 2gT^2/\pi^2$. The parametric equation of such a cycloid is given by

A1.6 $$\begin{cases} x = (gT^2/\pi^2)(\theta + \sin\theta) , \\ y = (gT^2/\pi^2)(1 - \cos\theta) = (2gT^2/\pi^2)\sin^2(\theta/2) , \end{cases}$$

from which we find

A1.7 $$s = (4gT^2/\pi^2)\sin(\theta/2) = (2T/\pi)(2gy)^{\frac{1}{2}}$$

as required. Thus, on releasing the particle from rest at any point of the arc of this cycloid, the time of descent of the particle to 0 will be T.

APPENDIX 2

From (3.8) and (3.9) we have $K(s_0,s) = \sin(\theta-\alpha)/r$, where $r, \theta, \alpha,$ are shown in Figure 2. We define

A2.1 $$K(s_0,s_0) = \lim_{s\to s_0} \frac{\sin(\theta-\alpha)}{r} .$$

Now, from the geometry of Figure 2 we can show that

A2.2 $$\frac{dr}{ds} = \cos(\theta-\alpha) \quad \text{and} \quad \frac{d\alpha}{ds} = \frac{\sin(\theta-\alpha)}{r} .$$

From (A2.1) we find on using l'Hôpital's rule that

$$K(s_0,s_0) = \lim_{s\to s_0} \left[\frac{\cos(\theta-\alpha)(\frac{d\theta}{ds} - \frac{d\alpha}{ds})}{\frac{dr}{ds}} \right]$$

A2.3 $$= \kappa_0 - \lim_{s\to s_0} \frac{d\alpha}{ds} = \kappa_0 - K(s_0,s_0) ,$$

where κ_0 denotes the curvature of the curve C at the point P_0. It follows immediately that $K(s_0,s_0) = \kappa_0/2$ which exists provided C has finite curvature at the point P_0.

From (3.13) we define $k(\theta,\theta_0)$ by

A2.4 $$k(\theta,\theta_0) = \frac{\cos(\theta-\alpha)}{\kappa r} - \frac{1}{2}\cot((\theta-\theta_0)/2) ,$$

and we want to show that $k(\theta_0,\theta_0) = \lim_{\theta\to\theta_0} k(\theta,\theta_0)$ exists. It then follows that k is a continuous function of both θ and θ_0 on $[0, 2\pi] \times [0, 2\pi]$. Let us write

A2.5 $$\begin{cases} k(\theta,\theta_0) = N(\theta,\theta_0)/D(\theta,\theta_0), \quad \text{where} \\ N(\theta,\theta_0) = 2\cos(\theta-\alpha)\sin((\theta-\theta_0)/2) - \kappa r\cos((\theta-\theta_0)/2) , \\ D(\theta,\theta_0) = 2\kappa r\sin((\theta-\theta_0)/2) . \end{cases}$$

Since, as $P \to P_0$, $r \to 0$ and both θ and α tend to θ_0, we have $N(\theta_0,\theta_0) = D(\theta_0,\theta_0) = 0$. Again, since $(\frac{dr}{ds})_{\theta=\theta_0} = 1$, we find $N'(\theta_0,\theta_0) = D'(\theta_0,\theta_0) = 0$, where prime denotes differentiation with respect to θ. In order to evaluate $k(\theta_0,\theta_0)$ we apply l'Hôpital's rule twice, to give

A2.6 $$k(\theta_0,\theta_0) = N''(\theta_0,\theta_0)/D''(\theta_0,\theta_0) .$$

After a little algebra we find that $D''(\theta_0,\theta_0) = 2$ and

$$N''(\theta_0,\theta_0) = -2\kappa'(\theta_0)r'(\theta_0) - \kappa_0 r''(\theta_0) .$$

Now, $r'(\theta_0) = (\frac{dr}{ds} \cdot \frac{ds}{d\theta})_{\theta=\theta_0} = 1/\kappa_0$. Again, since (see (A2.2)) we have $r'(\theta) = \cos(\theta-\alpha)/\kappa$, it follows that $r''(\theta_0) = -\kappa'(\theta_0)/\kappa_0^2$. Putting these results together we find that

A2.7 $$k(\theta_0,\theta_0) = -\kappa_0'/(2\kappa_0) ,$$

where κ_0' denotes $(d\kappa/d\theta)_{\theta=\theta_0}$. Thus we see that provided $d\kappa/d\theta$ exists at each point of C, then $k(\theta_0,\theta_0)$ is defined and k is continuous on $[0,2\pi] \times [0,2\pi]$. In the particular case when C is a circle then $\kappa'(\theta) \equiv 0$ for all θ and we have that $k \equiv 0$.

APPLICATION AND SOLUTION OF CAUCHY SINGULAR INTEGRAL EQUATIONS

E.O. Tuck
Applied Mathematics Department, University of Adelaide,
Adelaide, South Australia

1. Introduction

Singular integral equations arise naturally from boundary-value problems of interest in many areas of science and engineering. In particular, whenever a Green's function or isolated singular solution can be determined, the boundary-value problem can be replaced by an integral equation over part or all of the boundary. If the original boundary-value problem is elliptic, the integral equation is of Fredholm type, and its kernel is in general singular because of the singularity of the Green's function.

When the boundary condition involves derivatives normal to a smooth boundary surface, the integral equation is of the second kind, with the unknown function appearing outside of the integral as well as inside. Since such equations are reasonably well understood, and respond well to straightforward numerical methods, it is not proposed to spend much time on them in this paper. However, there is a large and growing body of research on numerical solution of important practical problems in engineering by this boundary integral equation technique, and there will be occasions to refer to this class of problem in the sequel. A useful recent review is that by Christiansen and Hansen (1978).

First kind integral equations, in which the unknown function appears only inside the integral, can arise from Dirichlet boundary conditions, i.e., from conditions not involving normal derivatives. The kernel of the integral equation is, however, only weakly singular in such cases, for smooth boundary surfaces. For example, a two-dimensional Dirichlet problem reduces to a single-variable integral equation with a logarithmically-singular kernel. Such integral equations arise in some of the applications to follow, and can, if desired, be converted to Cauchy-singular equations by differentiation. However, the undifferentiated equation responds well to numerical attack, and there is little to be achieved by such a conversion.

First kind Cauchy singular equations are the main topic of this paper, and these equations arise most naturally and directly from boundary-value problems for special non-smooth boundaries described (according to the context) as cuts, cracks, arcs, strips, ribbons, foils, slits, etc. Such boundaries enclose zero area or volume, and possess sharp edges, where singularities can be expected. Indeed, the strengths of these singularities may be output quantities of interest in applications.

The following is a rather selective survey of applications of such singular integral equations. Historically, there have been two dominant application areas, namely aerodynamics and elasticity, and my survey is biased in favour of the former. In fact, applications in elasticity will not be discussed, in spite of the fact that this is an area in which many published applications have appeared, and the best known book (Muskhelishvili, 1953) on singular integral equations is by an elastician. Samples of recent relevant work in the elasticity area are papers by Bogy (1975), Clements (1971), Wood and Head (1974), and Yokobori and Ichikawa (1967), and articles in Sih (1973). The recently translated monograph by Ivanov (1976) makes reference to applications in some other areas not covered by the present survey.

1.2 2D Thin Airfoil Theory

The theory of lifting thin wings in two and three dimensions grew along with the science and technology of aeronautics, and it is hard to pin down its originator, although Prandtl (1918) and Munk (1922) must be given much of the credit. The theory can be developed using techniques such as complex analysis, Fourier series, or separation of variables in elliptic coordinates, without explicit reduction to a singular integral equation, which also clouds the historical perspective from the point of view of the present paper.

In the two-dimensional incompressible case, the boundary-value problem is to solve Laplace's equation for $\phi(x,y)$ exterior to a thin wing, of upper surface $y = f_+(x)$ and lower surface $y = f_-(x)$, on which the normal derivative of $Ux + \phi$ vanishes, i.e.,

1.1 $$\frac{\partial\phi}{\partial y} = (U + \frac{\partial\phi}{\partial x})\, f_\pm'(x) \ , \quad \text{on} \quad y = f_\pm(x) \ , \quad |x| < \ell \ .$$

U is the free-stream velocity at infinity, where $\underset{\sim}{\nabla}\phi$ vanishes. This exact boundary condition is approximated, in the limit as $f_\pm \to 0$, by

1.2 $$\frac{\partial\phi}{\partial y} = Uf_\pm'(x) \ , \quad \text{on} \quad y = 0_\pm \ , \quad |x| < \ell \ .$$

The linearized problem so posed can be separated into its even and odd parts with respect to y, and we shall not be concerned further with the even part, which generates no lift. The odd part satisfies

1.3 $$\frac{\partial\phi}{\partial y} = Uf'(x) \ , \quad \text{on} \quad y = 0_\pm \ , \quad |x| < \ell \ ,$$

where

1.4 $$f(x) = \tfrac{1}{2}f_+(x) + \tfrac{1}{2}f_-(x)$$

defines the mean surface $y = f(x)$ of the wing. If

1.5 $$f'(x) = -\alpha = \text{constant}\ ,$$

the wing is uncambered and at angle of attack α; otherwise the variation in $f(x)$ defines the camber of the wing.

Solution of Laplace's equation subject to (1.3) may proceed in a number of ways. Here we simply choose a representation

1.6 $$\phi(x,y) = \frac{1}{2\pi}\int_{-\ell}^{\ell} d\xi\gamma(\xi)\ \arctan\frac{y}{x-\xi}$$

in terms of a distribution of vortices, and attempt to choose the vortex strength $\gamma(x)$ so that (1.3) is satisfied. This leads immediately to the singular integral equation

1.7 $$\frac{1}{\pi}\int_{-\ell}^{\ell}\frac{d\xi\ \gamma(\xi)}{x-\xi} = 2Uf'(x)$$

which is sometimes called the airfoil equation. The integration in (1.7) is to be interpreted in the Cauchy principal value sense.

1.3 General Solution of the Airfoil Equation

The airfoil equation (1.7) permits an exact solution. The aerodynamicists (e.g., Glauert, (1926)) originally solved it by expanding in a modified Fourier series, effectively a series in Chebyshev polynomials. This is still a powerful and effective numerical method, especially for generalizations of (1.7).

However, the solution to (1.7) can be written down directly as an integral, and we summarize results, e.g., as available from Muskhelishvili (1953) or Tricomi (1957) here for reference. If we write $g(x)$ for the righthand side $2Uf'(x)$, then one solution of

1.8 $$\frac{1}{\pi}\int_{-\ell}^{\ell}\frac{d\xi\ \gamma(\xi)}{x-\xi} = g(x)$$

is

1.9 $$\gamma_0(x) = -\frac{1}{\pi}\frac{1}{\sqrt{\ell^2-x^2}}\int_{-\ell}^{\ell}\frac{d\xi\ g(\xi)\sqrt{\ell^2-\xi^2}}{x-\xi}$$

But this solution is not unique, since any multiple of $(\ell^2-x^2)^{-\frac{1}{2}}$ satisfies the equation with $g(x) = 0$. Thus, the general solution of (2.8) is

1.10 $$\gamma(x) = \gamma_0(x) + \frac{\Gamma}{\pi\sqrt{\ell^2-x^2}} ,$$

where Γ is an arbitrary constant.

Note that $\int_{-\ell}^{\ell} \gamma_0(x)dx = 0$, so that

1.11 $$\int_{-\ell}^{\ell} \gamma(x)dx = \Gamma .$$

Thus, the arbitrary constant Γ has the aerodynamic significance of the net circulation around the wing, and the lack of uniqueness is simply a reflection of the fact that this circulation is arbitrary within irrotational flow theory. It should also be noted that as $x^2 + y^2 \to \infty$ in (1.6)

1.12 $$\phi(x,y) \to \frac{\Gamma}{2\pi} \arctan \frac{y}{x} + 0(\frac{1}{\sqrt{x^2+y^2}}) ,$$

so that the solution with $\Gamma = 0$ tends to zero at infinity, but that with $\Gamma \neq 0$ remains bounded. In a number of applications it is necessary that $\phi \to 0$ at infinity, and in that case $\gamma = \gamma_0(x)$ is the unique solution required; however, this is not true for airfoils.

An important feature of the solution $\gamma_0(x)$ is that, in general, it is infinite at *both* ends $x = \pm\ell$, behaving like the inverse square root of distance from these points. It is clearl possible to choose the arbitrary constant Γ so that the solution $\gamma(x)$ of (1.10) is non-singular at one, but not in general both, of these points. In particular, if

1.13 $$\Gamma = \int_{-\ell}^{\ell} d\xi\ g(\xi)\sqrt{\frac{\ell+\xi}{\ell-\xi}}$$

the solution $\gamma(x)$ is bounded (in fact, zero, vanishing like the square root of distance) at the trailing edge $x = +\ell$, and can be written

1.14 $$\gamma_1(x) = -\frac{1}{\pi}\sqrt{\frac{\ell-x}{\ell+x}}\int_{-\ell}^{\ell} \frac{d\xi\ g(\xi)}{x-\xi} \sqrt{\frac{\ell+\xi}{\ell-\xi}} .$$

In thin airfoil theory, boundedness or vanishing of γ at the trailing edge is normally demanded ("Kutta" condition), and hence $\gamma_1(x)$ is the required solution.

The solution $\gamma = \gamma_1(x)$ remains unbounded at the leading edge $x = -\ell$, unless

1.15 $$\int_{-\ell}^{\ell} \frac{d\xi\ g(\xi)}{\sqrt{\ell^2-\xi^2}} = 0$$

If $g(x)$ does satisfy the "orthogonality" condition (1.15), e.g., if it is an odd function of x, $\gamma = \gamma_1(x)$ vanishes at *both* ends $x = \ell$ and can be rewritten as

1.16 $$\gamma_2(x) = -\frac{1}{\pi}\sqrt{\ell^2-x^2}\int_{-\ell}^{\ell}\frac{d\xi\, g(\xi)}{x-\xi}\frac{1}{\sqrt{\ell^2-\xi^2}} .$$

The above exact results complete the solution for the simple airfoil equation (1.8), and the only *numerical* issue is that of efficient evaluation of the integrals involved. However, these solutions are also very useful in illustrating the qualitative properties of solutions of airfoil-like integral equations which do not have explicit analytic solution. Thus, whenever

1.17 $$K(x,\xi) = O\left(\frac{1}{x-\xi}\right) , \quad \xi \to x ,$$

the integral equation

1.18 $$\int_{-\ell}^{\ell} d\xi\, \gamma(\xi)\, K(x,\xi) = g(x)$$

possesses a family of solutions of the form

1.19 $$\gamma(x) = \gamma_0(x) + \Gamma\, \bar{\gamma}(x) ,$$

where $\gamma_0(x)$ is a particular solution, $\bar{\gamma}(x)$ is a non-trivial solution of the homogeneous equation, and Γ an arbitrary constant. In general, we must expect inverse square root singularities at both ends $x = \pm\ell$, one of which can always be eliminated by special choice of Γ. Both singularities can be eliminated only if $g(x)$ satisfies a suitable orthogonality condition.

1.4 3D Thin Wing Theory

Inclusion of the 3rd space dimension z in thin wing theory is a reasonably straightforwa extension, in principle. For wings that are *nearly* two-dimensional, i.e., of large span, Prandtl's (1918) lifting line theory provides three-dimensional corrections to the two-dimensional thin airfoil theory. Although I do not propose to discuss the basis for this theor it is of interest in the present context to note that Prandtl reduced the problem to an integro *differential* equation with a Cauchy singularity, of the form

1.20 $$\Gamma(z) = \Gamma_0(z) - \tfrac{1}{2}\ell(z)\int_{-s}^{s} d\zeta\, \frac{\Gamma'(\zeta)}{z-\zeta}$$

where $\Gamma_0(z)$ is the net circulation according to the two-dimensional theory, assumed known, an $2s$ is the span. Note that the chord 2ℓ could also vary with z.

In fact, as has been shown by Van Dyke (1964), Prandtl's lifting line equation (1.20) is inconsistent in order of magnitude, as an asymptotic expansion for large s/ℓ, and solution of

(1.20) gives a result formally no more accurate than that obtained by replacing Γ by Γ_0 under the integral sign. Nevertheless, (1.20) is interesting in its own right, and efficient numerical solution of (1.20) is a challenging problem.

If the span/chord ratio is *not* large, the problem is no longer even approximately two-dimensional. However, a fully three-dimensional theory for thin (nearly planar) wings proceeds in a quite analogous manner to that of §1.2, the line vortex distribution (1.6) being replaced by a so-called "horseshoe vortex". This leads to the lifting surface integral equation

1.21 $$\frac{1}{2\pi}\int_{-s}^{s} d\zeta \int_{-\ell(\zeta)}^{\ell(\zeta)} d\xi \, \frac{\gamma(\xi,\zeta)}{(z-\zeta)^2}\left[1 + \frac{x-\xi}{\sqrt{(x-\xi)^2 + (z-\zeta)^2}}\right] = 2Uf_x(x,z) \, .$$

Equation (1.21) is the direct generalization of the airfoil equation (1.7) to one extra dimension, and reduces to (1.7) if $\gamma(x,z)$ and $f(x,z)$ are independent of z. In contrast to (1.7) however, analytic solution is out of the question, and much recent effort has been devoted to development of efficient numerical schemes for solution of (1.21); for a survey of some competitive algorithms, see Wang (1974).

The kernel of (1.21) is exceedingly singular. For example, at fixed $x>\xi$ it possesses a *double* pole as a function of ζ, when $\zeta \to z$. On the other hand, for fixed $z \neq \zeta$, it appears not to be singular at all as $\xi \to x$, even though it reduces to the singular kernel of (1.7) in the large span limit. There now appears to be a large family of solutions, characterized by an arbitrary *function* $\Gamma(z)$, which is determined by applying a Kutta condition of boundedness of $\gamma(x,z)$ at *all* trailing points $x = +\ell(z)$. Further discussion of this interesting equation is perhaps beyond the scope of the present paper, but it is an indication that too much attention should not be concentrated on the relatively well understood case of integral equations in one dependent variable.

A single variable integral equation can arise in some special 3D problems. For example, if the dependence on the z-coordinate is sinusoidal, the problem can be reduced to an integral equation of the form (1.18), with a Bessel function kernel. This was done, Tuck (1974), for a "flapping wing" problem that also involves the extra complication of flow unsteadiness.

A similar reduction occurs in axisymmetric flow over an "annular airfoil", whose surface is everywhere close to the cylinder $y^2 + z^2 = a^2$. Obvious applications exist to shrouded propellors, engine nacelles, etc. The generalization of (1.7) now involves *ring* rather than line vortices, i.e., the annular airfoil is modelled as a distribution of "smoke rings", whose strength γ is to be determined. The resulting integral equation is of the form (1.18) with

1.22 $$K(x,\xi) = \frac{\partial}{\partial\xi}\frac{2}{\pi} m^{-\frac{1}{2}} \left[(1-\tfrac{1}{2}m) \underset{\sim}{K}(m) - \underset{\sim}{E}(m)\right]$$

$\underset{\sim}{K}(m)$ and $\underset{\sim}{E}(m)$ being complete elliptic integrals with parameter

1.23 $$m = \frac{4a^2}{(x-\xi)^2 + 4a^2} \, .$$

Note that as $\xi \to x$, $m \to 1$ and $\underset{\sim}{K}(m)$ becomes (logarithmically) singular; thus $K(x,\xi)$ has the property (1.17). Numerical solutions of integral equations with the kernel (1.22) have recently been used by King and Tuck (1978) for a ship hydrodynamic problem.

Finally, one should include in this category of 3D thin wing problems, the "low aspect ratio" limit, in which $s/\ell \to 0$, at the opposite extreme to Prandtl's lifting line theory (1.20). Although this theory can be developed as a formal limit of the lifting surface equation (1.21), it is perhaps more commonly thought of as a special case of slender-body theory.

The theory of slender bodies concerns elongated, but not necessarily nearly planar geometries, and the relevant approximation is to neglect variations with respect to the co-ordinate x along the direction of elongation. Thus, the 3D boundary value problem is reduced to a 2D boundary value problem in the cross flow (y,z) plane. For general non-planar bodies, one must solve this cross flow problem numerically, and in a number of application areas, efficient boundary integral equation methods have been used.

However, if the body is not only slender, but also nearly planar, i.e., has its y-coordinate everywhere small, the 2D cross-flow problem reduces to a problem for a strip $|z| < s(x)$, $y = 0$ in the (y,z) plane, and again, we can reduce the task to one of solving a Cauchy type singular integral equation. Indeed, the problem can be posed as one for the airfoil equation (1.8). In regions where the span $2s(x)$ is an increasing function of x, the appropriate solution is of the form (1.9), since non-zero net circulation Γ is not permitted. This leads to a simple theory for lift of low aspect ratio wings, pioneered by Jones (1946).

There are however, somewhat more general problems of this nature where the net cross flow circulation $\Gamma = \Gamma(x)$ at station x is not zero, and must be determined by a marching process as x increases. This leads (see, for example, Tuck (1979)) to a singular integral equation of *Volterra* type for the unknown $\Gamma(x)$, analogous to Abel's equation.

There is considerable literature on 3D Neumann boundary value problems for non-thin aerodynamic bodies. The first studies of this type of problem (see, for example, von Karman (1927)) were motivated by airship aerodynamics. The axisymmetric flow problem can be reduced to a single variable singular integral equation of the first kind by use of axial distributions of singularities, and a number of satisfactory methods (see, for example, Landweber (1951)) were developed for numerical solution of these integral equations prior to the advent of computers. However, once the computer age began, surface singularity methods (see, for example, Hess and Smith (1964)) soon proved able to solve fully three-dimensional boundary value problems. Hence, less attention has been paid to the simpler axial source method as a numerical tool, although it retains considerable interest in the context of the slender body approximation discussed above.

1.5 Viscous Fluid Problems

Most important viscous fluid problems are non-linear. However, if the velocity is sufficiently small, the problem can be linearized, leading to the equations for "creeping" or Stokes flow. In the steady 2D case the appropriate equation is the *biharmonic* equation for the stream function ψ and, as in the equivalent case of plane elasticity, a number of problems arise which can be reduced to singular integral equations, some of which possess exact closed form solutions (see, for example, Guiney (1972)).

A generalization is to *unsteady* 2D Stokes flow, described by

1.24 $$\nabla^4\psi = \alpha^2\nabla^2\psi$$

where $\alpha^2 = i\sigma/\nu$, σ being the frequency of oscillation and ν the viscosity. The appropriate Green's function is

1.25 $$\psi = \Psi(R) = \frac{1}{2\pi\alpha^2}\,[\log R + K_0(\alpha R)]$$

where R is distance from the singularity, and K_0 a Bessel function. Note that $\Psi = O(R^2 \log R)$ as $R \to 0$, so that a Cauchy type of singularity appears only after 3 differentiations.

The Green's function (1.25) may be used to reduce problems involving oscillations of or past ribbons or strips, to singular integral equations. For example, oscillations of the strip $y = 0$, $|x| < \ell$ normal to itself can be reduced to (1.8) with kernel

1.26 $$K(x,\xi) = \Psi''(|x-\xi|)\ .$$

The unknown $\gamma(x)$ is the pressure distribution over the strip, and the given function $g(x)$ is the normal velocity. Since only two differentiations of Ψ have been performed, the above K does not satisfy (1.17); however, a Cauchy type integral equation results from one further differentiation of both sides with respect to x. Results of numerical solution of this integral equation are contained in Tuck (1969).

A somewhat simpler flow is the 3D flow produced by infinitely long cylinders oscillating along their own axis. This means that only one velocity component, say w, in the z-direction, is excited by tangential stresses at the surface, and must be determined as a function of x and y. The case when this cylinder is a flat ribbon was examined by Levine (1957), who reduced the problem to an integral equation of the form (1.8) with

1.27 $$K(x,\xi) = K_0(\alpha|x-\xi|)\ .$$

Again, this K has a logarithmic, rather than Cauchy, singularity.

Problems involving slits or gratings in walls are in a sense, complementary to those involving ribbons or strips, and similar techniques can be used to reduce them to singular integral equations. Macaskill (1977) treated a single slit in a fixed wall in this manner. The kernel of Macaskill's equation is expressed in terms of an integral of Bessel functions, and in the form in which it was used, possesses a *double*-pole singularity. With care, this type of equation can be integrated once on both sides to reduce back to a Cauchy-type equation.

Problems involving arrays of slits were considered by Macaskill and Tuck (1977). The application here is to a screen or porous sheet. If the pores of the screen are small enough, the local flow through them may be slow enough to justify the "creeping" assumption, even when the flow at a large distance from the screen is sensible. Hence there is a potential relevance of this work to computation of important engineering parameters, such as acoustic permeability of materials. Again, the integral equation used for the numerical computations presented by Macaskill and Tuck possesses a double-pole singularity.

1.6 Heat Conduction Problems

Steady flow of heat is described by Laplace's equation for the temperature field T, and hence there are a number of such problems which can be reduced to airfoil-like integral equations. Different types of boundary conditions apply however, and the solutions required are normally of the type (1.9) rather than (1.14).

A problem treated by the present author (Tuck (1972)) involved design of thin disc-like bodies intended to measure net heat flux, e.g., in soil-science applications. The meter appears to the temperature field as a discontinuity in the coefficient κ of heat conduction or diffusivity. If the meter is thin, this discontinuity provides a temperature jump across the limiting plane, analogous to the vortex strength $\gamma(x)$ in (1.7). In the 2D case, the resulting analysis yields an integro-differential equation of the form of the Prandtl lifting line equation (1.20), where the given function Γ_0 is related to the meter diffusivity and thickness distributions.

The corresponding axisymmetric problem is of greater practical significance, since actual heat flux meters are normally disc-shaped. The analysis of such discs proceeds somewhat like the annular airfoil described in §1.4, the fundamental singularity being a ring source like (1.22). The resulting integro-differential equation does not seem to have been studied, either analytically or numerically.

If the temperature field is time dependent, it is no longer described by Laplace's equation and the equivalence with airfoil problems is not complete. For example, if the time dependence is pure sinusoidal at frequency σ, the governing equation is the same as for the axial viscous fluid oscillations discussed in §1.5, with $\alpha^2 = i\sigma/\kappa$, and the Green's function is a multiple of $K_0(\alpha R)$. Thus, the only change from the case when $\sigma = 0$ involves replacing, for example, the kernel $\frac{1}{x-\xi}$ of (1.8) by $\frac{\partial}{\partial\xi} K_0(\alpha|x-\xi|)$.

1.7 Acoustic and Electromagnetic Problems

Boundary integral equation methods for numerical solution of diffraction or scattering by arbitrary bodies have been well developed in acoustics (see, for example, Copley (1968)) and electromagnetism (see, for example, articles in Mittra (1973)). Much work has also been done on the special case of flattened bodies, such as ribbons and strips (see, for example, Hönl,

Maue and Westphal, (1961)). Of course, the complementary problem of a slit in a wall is, at least in the short-wave limit, the prototype diffraction problem, and hence is of great importance in applications, e.g., to optics.

The 2D scalar diffraction problem for a straight ribbon can be solved formally as a series of Mathieu functions (see, for example, Morse and Rubenstein (1938)). However, numerical values for the scattered field are perhaps more easily obtained by conversion to a singular integral equation, and direct numerical solution of that equation. Numerical results were presented by Taylor (1971 ; see also Tuck 1970), for a variety of boundary conditions.

The basic problem for normally incident plane waves of wave number k is to solve Helmholtz's equation

1.28 $$\nabla^2\phi + k^2\phi = 0$$

for the scattered field ϕ exterior to the strip $y = 0_\pm$, $|x| < \ell$, on which the mixed boundary condition

1.29 $$\phi_y(x,0) = 1 \pm P(x)\phi(x,0_\pm)$$

is to be satisfied. The function $P(x)$ is the permeability of the ribbon; if $P = 0$ it is "hard" or impermeable and the boundary condition is of Neumann type. In the limit $P \to \infty$, the ribbon becomes soft, and the boundary condition is of Dirichlet type. The normal velocity ϕ_y is assumed continuous across the ribbon, but there is a jump in the potential ϕ itself given by (1.29).

The problem can be solved by representation in terms of the Green's function

1.30 $$\phi = -\frac{1}{4}\, i\, H_0^{(1)}(kR) ,$$

where $H_0^{(1)}$ is a Hankel function, and Taylor (1971) reduces it to the integral equation

1.31 $$\frac{-i}{2}\int_{-\ell}^{\ell} \gamma(\xi)H_0^{(1)}(k|x-\xi|)d\xi + \frac{1}{k}\int_0^x P(\xi)\gamma(\xi)\sin k(x-\xi)\, d\xi = -k^{-2} + A\cos kx + B\sin kx$$

for $\gamma(x) = \phi(x,0_+)$, where A,B are arbitrary constants. These constants are ultimately determined by requiring boundedness of γ at the ends $x = \pm\ell$. Equation (1.31) is remarkable not only for the appearance of these arbitrary constants, but also for its mixed Fredholm-Volterra character.

The Neumann problem with $P(x) \equiv 0$ gives a logarithmically singular 1st kind Fredholm integral equation, which can be differentiated if necessary to give a Cauchy singularity. The novel feature of this equation is that the requirement that its solution be bounded at *both* ends $x = \pm\ell$ means that the given function on the right must satisfy a suitable orthogonality condition equivalent to (1.15). This can provide one of the two equations needed to determine

A and B. However, Taylor (1971) proceeds directly with the undifferentiated form (1.31), and finds A,B by direct application of the boundedness conditions at $x = \pm \ell$.

1.8 Water-Wave Problems

Some water-wave problems are equivalent to the scalar acoustic-type problems of the previous section, and indeed Taylor's (1971) work was done for the purpose of solving a ship-hydrodynamic problem, in which the permeability function $P(x)$ measures the flux of incident wave energy *underneath* a ship, subject to beam-sea excitation in shallow water.

If the water is not shallow, the unique surface character of water waves manifests itself, and special techniques are needed. An example is the extensive numerical work of Frank (1967) on computation of added mass and damping coefficients for oscillating cylinders. For arbitrary cylinder cross-section, Frank distributes wave sources over the cross section, the strength of which is determined by numerical solution of the resulting 2nd kind singular Fredholm equation. The Green's function G or wave-source potential must satisfy the free surface boundary condition and is in general quite a complicated function. In two dimensions G can be expressed in terms of the exponential integral function, and in three dimensions as a definite integral involving the exponential integral function in the integrand.

First kind integral equations of the type of interest for the present paper arise if the cylinder reduces to a flat plate, usually horizontal. If the plate is at non-zero submergence, the problem is qualitatively very similar to a thin-airfoil problem, the only difference being a modification to the kernel, replacing $\frac{1}{x-\xi}$ in (1.7) by $\partial G/\partial \xi$, and there are studies (see, for example, Keldysh & Lavrentiev (1937) and Siew & Hurley (1977)) of such problems, for application to hydrofoils or submerged breakwaters.

Even at zero submergence, the fundamental character of the resulting singular integral equation remains like that of the airfoil equation. Oscillating problems of this nature include the so called "finite dock" problem (see, for example, MacCamy (1961) and Holford (1964)). In the case of steady flow, this class of problem has application to *planing surfaces*, i.e., to surface vehicles of small draft.

The planing surface problem has a long history and a considerable body of literature has built up on both the 2D and 3D problems. Typical recent papers include Doctors (1977) and Tuck (1975). In the limit as free surface effects disappear, which occurs as the boat's speed gets higher and higher, the problem for a given wetted hull surface is in fact *identical* to that for the lower surface of a thin wing, and the corresponding aerodynamic results can be borrowed without change.

However, even so, there is a uniqueness difficulty still to overcome, since for application to real planing surfaces, it is not clear that we ought to be prescribing the wetted surface in advance. Thus, the true 2D high speed planing problem reduces to integral equations like (1.7), but with the range 2ℓ of integration to be determined, or in 3D to (1.21) with $\ell(z)$ to be determined. The conditions ultimately determining ℓ are continuity of body surface and free surface at their points of contact. Integral equations over domains whose extent is to be determined as part of the solution to the problem do not seem to have been studied, and it appears that one must adopt an inverse approach, fixing the wetted extent and allowing the solution to determine, in part, the hull shape giving such a wetted surface. This approach has recently been pursued by Oertel (1975) and Casling (1978).

Certain vertical boundary water wave problems can also be reduced to solution of Cauchy type integral equations. For example, the problem of water wave reflection and transmission by one or more holes or slits in an otherwise impermeable vertical wall or breakwater, comes into this category. The resulting integral equation permits a closed form solution in some cases; see, for example, Mei (1966), Guiney (1972), Porter (1972). However, it is perhaps better solved by direct numerical methods, as has been done by Macaskill (1977).

2. Numerical Solution

2.1 Why Integral Equations?

Since the applications discussed in Part 1 have a natural formulation as a boundary value problem for a partial differential equation, it is not entirely obvious that preliminary conversion to an integral equation is desirable. The question is whether direct discretization of the boundary value problem in space, e.g., using finite difference or finite element techniques is preferable to discretization of the equivalent integral equation over the boundary.

A quick, but not entirely satisfactory, answer is that the boundary integral equation technique is preferable because it reduces the effective number of dimensions of the problem. That is, any "space discretization" method requires determination of unknown function values over the *whole* of the region of interest, whereas the integral equation method only requires unknowns spread over the *boundaries* of that region. Indeed, normally one needs only distributions over the finite portions of that boundary, so the comparison is particularly favourable to the integral equation method for unbounded regions.

To fix ideas, suppose that N is a measure of the number of boundary points needed to provide adequate accuracy in any given problem. Typically, if the original boundary value problem is two-dimensional, $N = 50\text{-}200$ reasonably defines the boundary curve, and allows solutions with 3-4 figure accuracy, whereas if the problem is three-dimensional, $N = 500\text{-}2000$ may be necessary. Then, to achieve comparable accuracy, a space discretization will require about N^2 (say 10^4) unknowns in 2D, or $N^{3/2}$ (say 10^5) in 3D. These orders of unknowns are at or beyond feasible limits, so that in general one would have to compromise on accuracy to use space discretization.

The above is, however, a too simplified view of the problem. First of all, it is necessary that reduction to a boundary integral equation be *possible*. In most of the applications discussed in Part 1, this is certainly true. However, one important situation where it is *not* true involves *inhomogeneous* media. There, space discretization techniques, especially the finite element method, come into their own (indeed are almost obligatory), since no computationally acceptable algorithm for generating the necessary kernel or Green's function is likely for a general (perhaps numerically specified) degree of inhomogeneity. On the other hand, *non-linearity* of the governing equation is *not*, per se, a reason for preferring space discretization, since whatever method is chosen will require iteration of linear problems at some stage, and it is immaterial whether this is done on the boundaries or over the whole space, providing the medium is homogeneous.

The other important consideration in this comparison is the *sparseness* of the matrix of influence coefficients resulting from space discretization. Numerical approximation to the derivatives of a function at a given point involves only immediately neighbouring function values, and this leads to influence matrices which have many zero entries. On the other hand, a Fredholm integral equation, by its very nature, connects every function value with every other, so that there are no zero elements at all.

Thus, the time-to-compute comparison depends crucially on the extent to which the sparse character of the space discretization matrix can be exploited. As a rough guide, if it is assumed that the "time to invert" an $N \times N$ matrix by direct methods, e.g., Gaussian elimination, is $0(N^3)$, then this is a measure of the time needed for boundary integral equation methods. The corresponding matrix for a 2D space discretization is $N^2 \times N^2$, and is competitive only if a *sparse* $M \times M$ matrix can be inverted in a time of $0(M^{3/2})$. The corresponding 3D case requires a time $0(M^2)$. These times are probably achievable with current techniques, so that the time-to-compute argument does not necessarily favour the boundary integral equation technique, in spite of its smaller number of unknowns.

The above is still too superficial a comparison, as many important factors have been ignored. For example, although the boundary integral equation matrix is not sparse, it is normally strongly diagonally dominant, and in some circumstances its inversion may also be speeded up by use of the property that many of its elements are, if not actually zero, at least small. Physically, this simply reflects the fact that distant facets of a surface affect each other much less than closely neighbouring facets.

Another important consideration is of course storage limitation. Keeping N below one or two hundreds, enables retention of all elements in the computer's main memory; the alternative is frequent data transfer in and out of core, thereby incurring time and cost penalties. Thus, the boundary integral equation method seems clearly superior for problems in which such low N values give acceptable accuracy, and this includes almost all 2D problems.

With so many factors influencing efficiency of computation (even if that quantity could be satisfactorily defined!), there seems little point in a dogmatic preference for one or the other. As in most computational practice, much depends on the skill and even the taste of the programmer. Nevertheless, it would be fair to say that the boundary integral equation method is presently preferred by most researchers and practitioners of the art of computation, in the areas of application discussed in Part 1. What few comparative studies have been made (see, for example, Bai & Young (1974)) seem to favour it.

The paper by Bai & Young is in fact an example of a rather desirable class of studies which attempt to combine the best features of space and boundary discretization. Other recent references of this nature include Mei (1978), Zienkiewicz et al. (1977), Carson & Cambrell (1972) and Jami & Lenoir (1977). A recent article by Parlett (1978) raises issues of relevance to the topic of the present section.

2.2 General Boundary Integral Equation Discretization

For a general boundary surface, boundary value problems of interest in applications normally reduce to singular Fredholm integral equations of the second kind. In some cases, iterative solution commencing with neglect of the integral part may converge, and this may be used as a practical numerical technique. The only numerical issue is then adequate evaluation of the requisite integrations.

However, a more natural approach, almost universally adopted, is discretization of the integral, followed by "matrix inversion", i.e., solution of a set of linear algebraic equations. Of course, if one uses an iterative technique based on extraction of the diagonal elements of the matrix to solve this system, the net effect is no different from what is achieved by iteration with the original integral equation. Thus, we confine attention to discretization followed by non-iterative inversion.

There are an unlimited number of possible methods of discretization. One important method, used especially in aerodynamic applications, involves substitution of a suitable truncated series for the unknown function, the coefficients of which form the set of algebraic unknowns. This procedure has considerable advantages, if the set of basis functions can be chosen so that they match the expected properties of the solution well. In that case, the series may be expected to converge rapidly, and truncation to a small number of unknowns may yield adequate accuracy. For example, the lifting surface equation (1.21) has been solved satisfactorily for some aerodynamic purposes with as few as 5 or 10 terms of such series in each dimension (see, for example, Wang (1974)). However, even for lifting surfaces, this approach is too crude for some applications (see, for example, Cheng (1978)).

As soon as accuracy demands, or lack of suitable intuitive choices of basis functions, require unknown numbers larger than 50 or so, the above advantages of series substitution seems outweighed by (among other things) its lack of direct equivalence to the original problem. Thus if one retains actual function values as unknowns, the resulting matrix is intuitively understood as expressing the influence of one element or another.

It is also true that direct discretization normally leads to matrix elements that are easier to evaluate than those arising from series substitution. Here, as in many similar situations, there is a trade off between the time to *evaluate* and the time to *invert* the matrix. It is my personal (somewhat mystical!) viewpoint that such computations are most efficient when these two times are roughly *equal*. Since it is not until N takes values of the order of 50 that most inversion routines really cost $O(N^3)$ times, this idea carries with it the corollary that one really shouldn't waste one's time with discretizations involving less than 50 points.

Even if one confines attention to direct discretization of the integral, with function values as unknowns, there remain infinitely many ways to do this, corresponding to all possible numerical quadrature formulae. The singularity in the integral demands some care, but this is less of a problem than may appear.

In particular, the very definition of Cauchy principle value integrals suggests that, when the singularity is of such a nature, no numerical difficulty will be experienced. Indeed, in many cases (see, for example, Monacella (1967), Davis & Rabinowitz (1975)) one may simply proceed as if there were no singularity at all. Alternatively, the obvious subtraction procedure

2.1 $$\int_{-\ell}^{\ell} \gamma(\xi)K(x,\xi)d\xi = \int_{-\ell}^{\ell} (\gamma(\xi)-\gamma(x))K(x,\xi)d\xi + \gamma(x)\int_{-\ell}^{\ell} K(x,\xi)d\xi$$

eliminates the singularity in systems of the form (1.18) and is computationally acceptable whenever a routine exists for the ξ integral of the kernel K, which we assume to be the case.

Discretization can thus proceed according to whatever order of error is desired. Use of a low order quadrature, such as the midpoint or trapezoidal rule, gives a simple algorithm, but may require a larger value of N than a higher order quadrature, e.g., Simpson's rule or better. I make no attempt here to survey the variety of choices that have been tried (but see, for example, Hess (1975)), since again the argument is often no more than one of personal taste. However, it is clear that in a number of the application areas of interest, very low order discretizations have been used with success, and from now on we restrict attention to use of approximations equivalent to the midpoint rule.

2.3 STEP FUNCTION DISCRETIZATION

We use as our example, discretization of the integral equation (1.18), and dissect the interval $(-\ell,\ell)$ into segments (ξ_j,ξ_{j+1}), $j = 1,2,\ldots,N$ which need not be of uniform length. As perhaps the most elementary form of discretization, we first approximate the integrand as a step function, i.e., replace it on the jth interval by a constant, giving

2.2 $$\int_{-\ell}^{\ell} \gamma(\xi)K(x,\xi)d\xi \simeq \sum_{j=1}^{N} \gamma(x_j)K(x,x_j)(\xi_{j+1}-\xi_j)$$

where $x_j = \frac{1}{2}\xi_j + \frac{1}{2}\xi_{j+1}$ is some point in the jth interval. If we now evaluate the integral at $x = x_i$ for $i = 1,2,\ldots,N$ the integral equation (1,8) reduces to the set of linear equations

2.3 $$\sum_{j=1}^{N} K_{ij}\,\gamma_j = g_i$$

where

2.4 $$\gamma_j = \gamma(x_j)$$

2.5 $$g_i = g(x_i)$$

and

2.6 $$K_{ij} = K(x_i,x_j)\ (\xi_{j+1}-\xi_j)\ .$$

The above discretization applies in the first instance only to non-singular kernels $K(x,\xi)$. However, if K has the singular property (1.17) and x_j is the midpoint of the jth interval, the Cauchy interpretation of the integral implies simply that $K_{ii} = 0$, the off diagonal elements being still given by (2.6).

The error in the above discretization is expected to be proportional to some measure of the smoothness of the *complete* integrand γK. If K is singular, one may have doubts as to the legitimacy of such a procedure, especially for the diagonal and near diagonal elements. A slight modification of the above procedure has often been made to circumvent this difficulty, by requiring only that the function $\gamma(\xi)$ itself be approximable as a step function, rather than its product with K.

Thus, if we assume that, on the jth interval (ξ_j,ξ_{j+1}), $\gamma(\xi) = \gamma_j$ = constant, we arrive again at the system (2.3), but where now instead of (2.6), we have

2.7 $$K_{ij} = \int_{\xi_j}^{\xi_{j+1}} K(x_i,\xi)d\xi\ .$$

Equation (2.7) is an obvious generalization of (2.6) and reduces to (2.6) if the midpoint rule is used to evaluate the integral in it. But (2.7) does not in any way require K to be well behaved, and hence is preferable to (2.6) for kernels K which are singular, or rapidly oscillating, etc. Note that (2.7) applies even for $i = j$.

The apparent disadvantage of (2.7) over (2.6) is that it requires a knowledge, not of the kernel itself, but rather its integral. For example, if $K^{(\xi)}(x,\xi)$ is any function such that

2.8 $$\frac{\partial K^{(\xi)}}{\partial \xi}(x,\xi) = K(x,\xi)\ ,$$

then

2.9 $$K_{ij} = K^{(\xi)}(x_i,\xi_{j+1}) - K^{(\xi)}(x_i,\xi_j)\ .$$

In fact, this is often no disadvantage at all, since for most problems the anti-derivative $K^{(\xi)}$ is as easy to compute as the original function K. For example, routines for integrals of Bessel functions are no harder to program or use than corresponding routines for the Bessel functions themselves.

However, should this not be the case, it is neither necessary nor desirable to evaluate the integral in (2.7) more accurately than is commensurate with the approximations already made. For example, (King & Tuck (1979)) when K has the property (1.17), a suitable procedure is to subtract off the singular part and integrate that exactly, using the midpoint rule on the remainder. Thus, if

2.10 $$K(x,\xi) = \frac{1}{x-\xi} + \bar{K}(x,\xi) \ ,$$

then

2.11 $$K_{ij} = \log \left|\frac{x_i-\xi_j}{x_i-\xi_{j+1}}\right| + \bar{K}(x_i,x_j)(\xi_{j+1}-\xi_j) \ .$$

It is also true that, the further from the diagonal, the less significant the improvement of (2.7) over (2.6). This has a clear physical interpretation, exploited in perhaps the best-known algorithm of this type, the "Douglas Program" of Hess & Smith (1964). In that program a three dimensional surface is approximated by a collection of quadrilateral panels, on each of which the problem is solved by assuming uniform source strength. This still leaves a somewhat unpleasant integration task, the equivalent of (2.7) (but in two variables), and considerable savings in computer time are achieved by use of (2.6) instead, when j is sufficiently different from i. The physical equivalence is between distributed sources over the quadrilateral, and a single point source of the same total strength, located at the centroid of the quadrilateral. These two quantities are indistinguishable at a sufficient distance.

2.4 Second Kind Equations

Although our main concern here is with first kind equations of the form (1.18), the corresponding second kind equation

2.12 $$\gamma(x) + \int_{-\ell}^{\ell} d\xi\gamma(\xi)K(x,\xi) = g(x)$$

arises in many applications and deserves comment. To a certain extent the distinction between first and second kind integral equations is blurred for singular kernels, since (2.12) is simply obtained by the replacement

2.13 $$K(x,\xi) \rightarrow \delta(x-\xi) + K(x,\xi)$$

in (1.18). That is, second kind equations are just first kind equations with δ-function singularities in the kernel.

The corresponding discretized system of equations replacing (2.3) is

2.14 $$\gamma_i + \sum_{j=1}^{N} K_{ij}\,\gamma_j = g_i \ .$$

In the case when K has the Cauchy property (1.17), so that K_{ii} is either zero (according to (2.6)) or small (according to (2.7) , the term in γ_i is responsible for a diagonally dominant character to the governing matrix $[\delta_{ij} + K_{ij}]$.

Thus, in general, we expect no troubles in inversion of (2.14) to determine γ_i, and there is ample computational experience to confirm this. Among many references which provide actual results, computer times, etc. for this class of algorithm, we mention Hess & Smith (1964), Frank (1967), Chen & Schweikert (1963), Richmond (1963), Tuck (1971), Taylor (1973), and Dawson (1977).

One minor difficulty occurs for some wave-like problems, in which the matrix $[\delta_{ij} + K_{ij}]$ becomes singular at certain eigenfrequencies. For example, for exterior Neumann problems, these frequencies are the eigenfrequencies of the corresponding *interior Dirichlet* problem. This is a purely numerical difficulty, since these frequencies have no physical significance, and the solution to the original exterior problem is known to exist at all frequencies. Various techniques (see, for example, Sayer & Ursell (1977)) have been suggested and used successfully to eliminate this difficulty.

It should also be observed that in wave-like problems, one may expect to have to reduce the mesh size to a value significantly less than the fundamental wavelength, in order to achieve acceptable accuracy. However, this may not always be the case, if the discretization corresponding to (2.7) is used, as distinct from that corresponding to (2.6). That is, the kernel K may have a rapidly varying character in some cases where the expected solution $\gamma(x)$ is not rapidly varying. Since (2.7) only approximates γ itself, in such cases one may expect satisfactory accuracy for wavelengths significantly less than the meshsize. This property is analogous to that of Filon's quadrature (see, for example, Tuck (1967)).

2.5 First Kind Cauchy Singular Equations

If the task is to solve (2.3), where K has the property (1.17), the matrix $[K_{ij}]$ to be inverted is far from diagonally dominant, and attempts at direct inversion fail. Indeed, they must fail, and we should be very unhappy if they succeeded, since this would imply a unique solution to an integral equation that is known not to possess such a unique solution! As a deliberate test of this idea, Macaskill (1977) attempted to invert K_{ij} for the ordinary airfoil equation, and obtained wildly oscillating large numbers for $[\gamma_i]$.

We should expect that, at least in the limit as $N \to \infty$, $[K_{ij}]$ is a singular matrix, since the homogeneous equation with $g(x) \equiv 0$ possesses a non-trivial solution. This suggests one possible solution procedure, involving extraction of the eigenvector corresponding to this non-singular matrix. However, the following method seems easier to implement, and works well.

First, we should observe that first kind integral equations with *logarithmically* singular kernels possess a unique solution. For example, if

2.15 $$K(x,\xi) = \log |x-\xi|$$

and $g(x) = 1$, the solution of (1.18) is

2.16 $$\gamma(x) = \frac{1}{\pi \log \frac{1}{2}\ell} (\ell^2 - x^2)^{-\frac{1}{2}} .$$

In fact, this is simply a (unique) multiple of the homogeneous solution of the airfoil equation (1.8), which results on differentiation with respect to x. Morland (1970) uses such reductions to provide semi-analytic solutions for some classes of logarithmically singular kernels.

Thus, if instead of the Cauchy property (1.17), we have a logarithmic character

2.17 $$K(x,\xi) = O(\log|x-\xi|) \ , \quad \xi \to x \ ,$$

we have no reason to believe that inversion of $[K_{ij}]$ will fail, and it does not. The diagonal elements K_{ij} computed from (2.7) are no longer small, and the matrix is satisfactorily diagonally dominant. For example, Macaskill (1977) gives a table showing rates of convergence toward the solution (2.16) as $N \to \infty$, for various choices of mesh (see later).

Suppose now that we have to solve the integral equation (1.18) with a Cauchy kernel K satisfying (1.17). Let $K^{(x)}$ be any function whose x-derivative is K, i.e.

2.18 $$\frac{\partial}{\partial x} K^{(x)}(x,\xi) = K(x,\xi) \ .$$

Then $K^{(x)}$ has the logarithmic property (2.17).

Now, integrate the original integral equation (1.18) once with respect to x, obtaining

2.19 $$\int_{-\ell}^{\ell} d\xi \ \gamma(\xi) \ K^{(x)}(x,\xi) = g^{(x)}(x) + \Gamma \ ,$$

$g^{(x)}$ being any antiderivative of g, and Γ an arbitrary constant.

From what we have observed about logarithmic kernels, when the righthand side of (2.19) is a *known* function, (2.19) possesses a unique solution, that can be obtained without trouble by discretization and matrix inversion. However, the righthand side of (2.19) is not entirely known. Any solution $\gamma(x)$ must be a linear function of the constant Γ, and we can write it in the form (1.19), where $\gamma_0(x)$ satisfies

2.20 $$\int_{-\ell}^{\ell} d\xi\gamma_0(\xi) \ K^{(x)}(x,\xi) = g^{(x)}(x)$$

and $\bar{\gamma}(x)$ satisfies

2.21 $$\int_{-\ell}^{\ell} d\xi\bar{\gamma}(\xi) \ K^{(x)}(x,\xi) = 1$$

for all x in $(-\ell,\ell)$. Both $\gamma_0(x)$ and $\bar{\gamma}(x)$ are uniquely determined, and thus we have provided a procedure for generation of the general solution (1.19), as required.

The matrix to be inverted involves use of $K^{(x)}$ instead of K, and hence, if (2.9) is to be used for its elements, we need a routine for the *double* antiderivative $K^{(x,\xi)}$ such that

$\partial^2 K^{(x,\xi)}/\partial x \partial \xi = K$. This, again, is not normally any more difficult to construct than routines for K itself. Once the matrix $[K_{ij}^{(x)}]$ is set up, the inversion process yields both solutions $[\gamma_j^0]$ and $[\bar{\gamma}_j]$, at little more cost than if there had been only one unknown vector.

2.6 Choice of Mesh

So far, nothing has been said about the manner of dissection of the interval $(-\ell,\ell)$ into N subintervals (ξ_j,ξ_{j+1}), and in principle any such subdivision with

2.22 $$-\ell = \xi_1 < \xi_2 < \xi_3 \ldots < \xi_N < \xi_{N+1} = +\ell$$

is acceptable. An equal interval dissection

2.23 $$\xi_j = -\ell + (j-1)\,\frac{2\ell}{N}$$

has obvious advantages of simplicity, and may often be the best choice.

In particular, if the kernel function $K(x,\xi)$ actually involves only the *difference* $(x-\xi)$, there is a significant computational advantage to use of equal intervals, since at most $2N$ values of the kernel need to be computed. That is, in evaluating the N^2 matrix elements K_{ij}, we can make use of the difference character of the kernel, providing the mesh in uniform, to compute one row by a shift of the elements in the row above it. This is a very important potential time saving, in cases where the function K (or its integrals $K^{(\xi)}, K^{(x,\xi)}$ etc.) is complicated and costly to compute, as is the case in a number of the applications discussed in Part 1. The saving is significant only for relatively low N values, where the $O(N^3)$ inversion time is not dominant over the $O(N^2)$ evaluation time, and is achievable *only* for difference kernels on an equally spaced mesh.

If these conditions are not met, one expects at least N^2 kernel evaluations, and no one mesh is superior to any other. The choice depends then on the expected degree of singularity of the solution function $\gamma(x)$. If we are using the simple step function approximation to $\gamma(x)$, leading to matrix elements given by (2.7), we should aim to choose a mesh spacing which makes this approximation as acceptable as possible. That is, we should make the interval length $(\xi_{j+1} - \xi_j)$ small where the expected solution $\gamma(x)$ is rapidly varying.

In general boundary integral equation problems, this choice is usually an intuitive one, in which the boundary is discretized a priori by the user, in accordance with his personal tastes and expectations. There are a number of quite interesting and deep issues here, perhaps a little beyond the scope of the present paper.

For example, (see Denny (1963)) if one is really only interested in behaviour of the solution over a small portion of the boundary, where a great deal of detail is needed, then obviously the idea is to put many mesh points in that portion, with comparatively sparse density elsewhere. Just how far can one go with this nonuniformity? It would appear possible to automate the allocation of mesh points by some computer adaptive process, but then how do we stop the computer from demanding excessive detail near sharply varying points that are not really of interest to the user?

For integral equations of airfoil type (1.18), one choice of non-uniform mesh does seem to have inherent advantages, namely a "Chebyshev" mesh given by

2.24 $$\xi_j = -\ell \cos (j-1) \frac{\pi}{N} .$$

This choice accumulates mesh points near the ends $x = \pm \ell$, which is obviously desirable in view of the expected singularity there.

Indeed, the extent of accumulation given by (2.24) is exactly right, as can be seen by the change of variable

2.25 $$\int_{-\ell}^{\ell} f(x)(\ell^2 - x^2)^{-\frac{1}{2}} dx = \int_0^{\pi} f(-\ell \cos \theta) d\theta .$$

Thus, an integrand with an inverse square root singularity with respect to x, is converted to a *nonsingular* integrand with respect to θ. The result of applying the midpoint rule to the lefthand side of (2.25) using the Chebyshev mesh (2.24) is exactly the same as that arising from application of a *uniform* mesh for $0 < \theta < \pi$ to the righthand side. Thus, the Chebyshev mesh (2.24) integrates functions with inverse square root singularities with *no less* accuracy than a uniform mesh integrates nonsingular functions.

In practice, this means that one can expect a faster rate of convergence as N increases, by use of the Chebyshev mesh, than by use of a uniform mesh. Comparative tests by Macaskill (1977) indicate improvements from $O(N^{-1})$ to $O(N^{-2})$ convergence rates in this way, for the airfoil equation, using the discretization corresponding to the matrix elements (2.7).

In fact, a further option is open to us, in that nothing in the discretization (2.7) demands that the collocation points x_i at which we force the original integral equation to hold, be midpoints of intervals. In the most general case, the set of points $\{x_i\}$ can be quite independent of the set $\{\xi_j\}$. However, it seems likely that points x_i within the i'th interval, are desirable, and points not too far from the midpoint are likely to perform best.

One choice other than the actual or Cartesian midpoint, is the "Chebyshev" midpoint

2.26 $$x_i = -\ell \cos (i-\tfrac{1}{2}) \frac{\pi}{N} ,$$

which corresponds to a midpoint with respect to θ after the change of variables (2.25) has been carried out. Although, for large N, the values of x_i given by (2.26) differ very little from the true interval midpoints, the "fine tuning" allowed by (2.26) can make a significant difference to the convergence rate, which Macaskill (1977) finds is now $O(N^{-3})$ for the airfoil equation. This rate of convergence is probably as high as we can expect from stepfunction approximations, so that it would appear that the choice (2.26) is optimum for airfoil type problems.

In some aerodynamic contexts, success has been achieved by very special choices of collocation points. For example (see James (1972), Yeung & Tan (1979)) it is possible to use a discretization even coarser than the step function method, in which distributed vortices are replaced by discrete isolated vortices. One then has freedom to choose both the location of this vortex and the collocation point for each mesh element. There appear to be special advantages to the choice which locates the vortex at $\frac{1}{4}\Delta x$ and the collocation point at $\frac{3}{4}\Delta x$ distant from the interval leading point, where Δx is the interval length. In particular, the Kutta condition appears to be satisfied automatically in this case. This discrete vortex approximation is also commonly used in vortex sheet roll up studies (see, for example, Fink and Soh (1978)).

For equations more general than the airfoil type, the choice of mesh and collocation points remains largely intuitive. Of course, if one expects algebraic singularities of known strengths other than "$-\frac{1}{2}$", one can easily design special meshes to accommodate the singularity However, one soon reaches a point where the programming effort makes further refinement seem not worthwhile, and the simple equal interval mesh retains appeal, even at the cost of a somewhat slower convergence rate.

2.7 Determination of Free Constants

The procedure outlined in §2.5 allows efficient computation of the *general* solution (1.19) of airfoil-like integral equations. What we have available upon inversion of the matrix $[K_{ij}(x)]$ is a pair of vectors $[\overset{0}{\gamma}_j]$, $[\bar{\gamma}_j]$, being the discrete representation of the particular solution $\gamma_0(x)$ and the homogeneous solution $\bar{\gamma}(x)$ of the original integral equation (1.18).

In any particular problem we shall require a technique for determining the constant Γ in the general solution (1.19). That is, we need to establish just what multiple of the homogeneou solution $\bar{\gamma}$ to add to γ_0. Each application demands and provides its own subsidiary condition to effect this determination. The most important such subsidiary condition is the Kutta condition of aerodynamics, which demands that the solution vanish at the trailing edge, i.e., $\gamma(\ell) = 0$.

Now, unless the righthand side function $g(x)$ is very special, it is unlikely that this condition will be satisfied by γ_0, and it is certainly not satisfied by $\bar{\gamma}$. Thus, we expect in general that as $x \to \ell$, both $\gamma = \gamma_0$ and $\gamma = \bar{\gamma}$ possess inverse square root singularities. In fact, it can be shown (see, for example, Tricomi (1957)) that when $g(x)$ is well behaved at $x=\ell$, so is $(\ell-x)^{\frac{1}{2}}\gamma(x)$, where $\gamma(x)$ is any solution, i.e.,

2.27
$$\gamma(x) \to A(\ell-x)^{-\frac{1}{2}} + B(\ell-x)^{\frac{1}{2}} + O(\ell-x)$$

for some constants A,B. Note that there is no term in $\gamma(x)$ of "constant" order; if it happens that $A = 0$, then $\gamma(\ell)$ is not only finite, but in fact *zero*.

The numerical results for both γ_0 and $\bar{\gamma}$ will confirm these properties, and we expect therefore, quite large numbers for the end elements γ_N^0, $\bar{\gamma}_N$. One rather crude way (see, for example, Tuck & Newman (1974)) to enforce the Kutta condition is simply to choose

2.28 $$\Gamma = - \gamma_N^0 / \bar{\gamma}_N ,$$

which forces γ to take a zero value on the N'th interval.

Determination of Γ by (2.28) must normally take place subsequent to determination of both solutions $[\gamma_j^0]$, $[\bar{\gamma}_j]$. However, a novel alternative approach, adopted by Oertel (1975), is to incorporate this requirement into the original system of linear equations. That is, without separate definition of γ^0, $\bar{\gamma}$, Oertel seeks a solution γ whose discrete approximation $[\gamma_j]$ possesses a zero last element $\gamma_N = 0$. In that case, the N'th column of the matrix $[K_{ij}^{(x)}]$ is superfluous, as is of course the N'th unknown γ_N. If we replace all $[K_{iN}^{(x)}]$ by "-1" and replace γ_N by Γ, the new N'th linear equation is equivalent to the Kutta condition $\gamma(\ell) = 0$, applied on the N'th interval. Thus, inversion of this modified matrix condition, solves the original problem subject to this subsidiary condition, with some saving of computer time over the procedure involving use of two unknown vectors.

Since the Kutta condition really should be applied at the end $x = \ell$ of the body rather than over the whole end element, a somewhat better procedure is to extrapolate the computed output in order to determine the parameter A in (2.27), and then choose Γ so as to cancel the inverse square root part of the solution. That is, if A^0, $\bar{A}$ are values of A for $\gamma^0, \bar{\gamma}$ respectively, we choose

2.29 $$\Gamma = - A^0/\bar{A}$$

instead of (2.28). However, if the term in B is deleted from (2.27) and A^0, $\bar{A}$ simply determined by fitting to the last computed elements γ_N^0, $\bar{\gamma}_N$, then (2.29) and (2.28) give identical results. Thus, improvement over (2.28) is obtainable only by including the term in B, and fitting (2.27) to the last two elements γ_{N-1} and γ_N. This procedure has been used successfully in Tuck (1974) and King & Tuck (1979).

Very similar procedures can be adopted in cases where the subsidiary condition is not of Kutta type. Thus, we may require the integral of γ to vanish, a property analogous tc the special solution (1.9) of the airfoil equation, which may correspond to $\phi \to 0$ at infinity in the equivalent boundary value problem. In that case, the constant Γ can be determined as

$$\Gamma = - \int_{-\ell}^{\ell} \gamma^0(x)\, dx \Big/ \int_{-\ell}^{\ell} \bar{\gamma}(x) dx .$$

Problems such as that corresponding to (1.31) may involve more than one arbitrary constant and hence demand more than one subsidiary condition, and these conditions may be incorporated either by Oertel's procedure, or by using as many righthand sides as there are arbitrary constants. As many as 6 real righthand sides were involved in the flapping wing computations in Tuck (1974).

One special feature of unsteady aerodynamic problems, such as that in Tuck (1974) is that the given righthand side function $g(x)$ is *not* well behaved at the trailing edge $x = \ell$, and hence (2.27) is not valid. In fact, this is necessary and desirable, since the modified Kutta condition for unsteady problems demands a finite *non-zero* value for $\gamma(\ell)$. Physically, $g(x)$ now contains a contribution from the unsteady vortex wake in $x > \ell$, and is singular when $x \to \ell$ in such a way as to produce a "constant" term between the inverse square root term (in A) and the square root term (in B) of (2.27). This property must be exploited numerically in carrying out the extrapolation to determine the unknown constants.

2.8 Singularities other than Cauchy Type

Although our main concern is with the simple pole kernel satisfying (1.17), a number of the applications mentioned in Part 1 involved other degrees of singularity. In particular, kernels with singularities corresponding to either integration or differentiation of Cauchy singular kernels occur frequently, and properties of the resulting integral equation can be inferred from those of the airfoil equation.

The case of one integration, leading to a logarithmic kernel as in (2.17), has already been discussed, and used as a tool for solving the Cauchy singular case. The essential property is that integral equations (1.18) with such logarithmic kernels possess a unique solution, for any righthand side function $g(x)$. It follows that integral equations (1.18) resulting from *one further* integration, i.e., whose kernels satisfy

2.30 $$K(x,\xi) = O((x-\xi)\log|x-\xi|\ , \quad \xi \to x\ ,$$

can be expected to possess *no solution at all*, for a prescribed function $g(x)$. For example, the integral equation

2.31 $$\int_{-\ell}^{\ell} d\xi\gamma(\xi)(x-\xi)\log|x-\xi| = 1$$

possesses no solution. On the other hand, if the righthand side is replaced by $x + C$, then a solution exists only if $C = 0$, namely

2.32 $$\gamma(x) = \frac{(\ell^2-x^2)^{-\frac{1}{2}}}{\pi(1+\log\frac{1}{2}\ell)}\ .$$

Kernels with the property (2.30) occur in some generalisations of the planing problems discussed in §1.8, and the physical interpretation of the lack of existence of solutions is not yet clear.

Another class of singular kernels results from differentiation, rather than integration of the Cauchy kernel. Thus, in particular, we may be interested in kernels with the property

2.33 $$K(x,\xi) = O(x-\xi)^{-2}\ , \quad \xi \to x\ .$$

Before proceeding, however, we need to clarify what is meant by the integral in (1.18), when K has the property (2.33), since a Cauchy principal-value interpretation is no longer appropriate. Indeed, the integral diverges, not only formally, but also according to most people's intuition, since it involves a one-signed infinite area near $\xi = x$. The appropriate interpretation is according to the "Hadamard principal value"

2.34 $$\int_{-\ell}^{\ell} d\xi\gamma(\xi)K(x,\xi) = \lim_{\varepsilon\to 0}\left[-\frac{2\gamma(x)}{\varepsilon} + \int_{-\ell}^{x-\varepsilon} + \int_{x+\varepsilon}^{\ell} d\xi\gamma(\xi)K(x,\xi)\right],$$

in a normalisation such that the constant of proportionality in (2.33) is unity. For positive $\gamma(x)$, the large negative constant $-2\gamma(x)/\varepsilon$ cancels the large positive contributions from the integral near $\xi = x$. It is also possible to reduce double pole integrals to simple pole integrals that have a Cauchy principal value interpretation by an integration by parts, but this is seldom computationally convenient.

One feature of double pole equations is that their solution is *doubly* non-unique. That is, the general solution must be expected to be of the form

2.35 $$\gamma(x) = \gamma^0(x) + C\bar{\gamma}(x) + D\bar{\bar{\gamma}}(x)$$

where γ^0 is a particular solution, $\bar{\gamma}$ and $\bar{\bar{\gamma}}$ are independent solutions of the homogeneous equation, and C,D arbitrary constants. For example, if $K = (x-\gamma)^{-2}$, $\bar{\gamma} = (\ell^2-x^2)^{-\frac{1}{2}}$ and $\gamma = x(\ell^2-x^2)^{-\frac{1}{2}}$, as can be seen by differentiation of the corresponding Cauchy singular equations.

Thus, in any problem where the kernel has such a double pole, we must be supplied with two subsidiary conditions. For example, (Macaskill & Tuck 1977) the unknown $\gamma(x)$ may be required to vanish at *both* ends $x = \pm\ell$. This particular set of subsidiary conditions is notable in that if the integration by parts procedure mentioned above is adopted, to convert to a Cauchy singular problem, the integrated part vanishes.

There seems no reason in principle why a procedure involving 3 righthand side vectors, quite analogous to that discussed earlier for numerical solution of Cauchy singular equations, should not enable determination of the general solution (2.35) in the present case, providing care is taken with numerical treatment of (2.34). However, Macaskill & Tuck (1977) adopted a rather more direct procedure, of discretization of the integrand according to (2.7) and (2.9), without paying any heed at all to the special singular character of K. The matrix $[K_{ij}]$ was then inverted, and the resulting solution $[\gamma_j]$ appeared to converge quite well as $N \to \infty$ to a result *which already satisfied the subsidiary conditions* $\gamma(\pm\ell) = 0$! A partial explanation for this fortunate event is provided in the paper by Macaskill & Tuck, but there still remain some interesting problems associated with such "super singular" kernels.

References

[1] Bai, K.J. and Yeun, R.W. Numerical solutions to free-surface flow problems, *10th Symp. Naval Hydro.,* Cambridge, Mass., Proc., 609-633, 1974.

[2] Bogy, D.R. The plane solution for joined dissimilar elastic semistrips under tension, *J. Appl. Mech.,* 42, 93-98, 1975.

[3] Carson, C.T. and Cambrell, G.K. The numerical solution of electromagnetic problems, *Symp. on Microwave Commun.,* Radio Research Board of Australia, University of Adelaide, 1972.

[4] Casling, E.M. Planing of a low-aspect-ratio flat ship at infinite Froude number, *J. Eng. Math.*, 12, 43-58, 1978.

[5] Chen, L.H. and Schweikert, D.G. Sound radiation from an arbitrary body, *J. Acoust. Soc. Am.*, 35, 1626-1632, 1963.

[6] Cheng, H.K. Lifting-line theory of oblique wings, *Univ. So. Calif.*, School of Eng. Report USCAE 135, July, 1978.

[7] Christiansen, S. and Hansen, E.B. Numerical solution of boundary-value problems through integral equations, *Z.A.M.M.*, 58, T14-T25, 1978.

[8] Clements, D.L. The response of an anisotropic elastic half-space to a rolling cylinder", *Proc. Camb. Phil. Soc.*, 70, 467-484, 1971.

[9] Copley, L.G. Fundamental results concerning integral representations in acoustic radiation, *J. Acoust. Soc. Am.*, 44, 28-32, 1968.

[10] Davis, P.J. and Rabinowitz, P. Numerical Integration, Blaisdell, 1967.

[11] Dawson, C.W. A practical computer method for solving ship-wave problems, *Proc. 2nd Int. Conf. Num. Ship. Hydro.*, University of California, Berkeley, 1977.

[12] Denny, S.B. Applicability of the Douglas company program to hull pressure problems, DTNSRDC Rep. 1786, Dept. Navy, Washington, D.C., 1963.

[13] Doctors, L.J. Theory of compliant planing surfaces, *Proc. 2nd Int. Conf.Num. Ship Hydro.*, University of California, Berkeley, 1977.

[14] Fink, P.T. and Soh, W.K. A new approach to roll-up calculations of vortex sheets, *Proc. Roy. Soc. Lond.*, A362, 195-209, 1978.

[15] Frank, W. On the oscillation of cylinders in or below the free surface of deep fluids, DTNSRDC Rep. 2375, Dept. Navy, Washington, D.C., 1963.

[16] Glauert, H. The Elements of Aerofoil and Airscrew Theory, Cambridge University Press, 1926.

[17] Guiney, D.C. Some Flows Through a Hole in a Wall in Viscous and Non-viscous Fluids, Ph.D. thesis, University of Adelaide, 1972.

[18] Hess, J.L. The use of higher-order surface singularity distributions to obtain improved potential-flow solutions for two-dimensional airfoils, *Comp. Math. Appl. Mech. & Eng.*, 5, 11-35, 1975.

[19] Hess, J.L. and Smith, A.M.O. Calculation of non-lifting potential flow about arbitrary three-dimensional bodies, *J. Ship Res.*, 8, 22-24, 1964.

[20] Holford, R.L. Short surface waves in the presence of a finite dock, I & II, *Proc. Camb. Phil. Soc.*, 60, 957-1011, 1964.

[21] Hönl, H., Maue, A.W. and Westphal, K. Theorie der Beugung, *Handb.Phys.*, 25, 218-573, 1961.

[22] Ivanov, V.V. The Theory of Approximate Methods and their Application to the Numerical Solution of Singular Integral Equations, Noordhoff, 1976.

[23] James, R.M. On the remarkable accuracy of the vortex lattice method, *Comp. Meth. Appl. Mech. Eng.*, 1, 59-79, 1972.

[24] Jami, A. and Lenoir, M. Formulation variationelle pour le couplage entre une methode d'elements finis et une representation integrale, *Comp. Rend. Acad. Sci. Paris*, Ser.A, 285, 269-272, 1977.

[25] Jones, R.T. Properties of low-aspect-ratio pointed wings at speeds below and above the speed of sound, NACA Rep. 835, 1946.

[26] von Karman, T. Calculation of pressure distribution on airship hulls, NACA, T.M. 574, 1930.

[27] Keldysh, M.V. and Lavrentiev, M.A. On the motion of a wing beneath the surface of a heavy fluid, *Proc. Conf. Theory Wave Res.*, Central Aero-Hydro. Inst., Moscow, 1937. (see Panchenkov, DTNSRDC Translation 316, 1963).

[28] King, G.W. and Tuck, E.O. Lateral forces on ships in steady motion parallel to banks or beaches, *Appl. Ocean Res.*, 1, 89-98, 1979.

[29] Landweber, L. The axially symmetric potential flow about elongated bodies of revolution, DTNSRDC Rep. 761, Dept. Navy, Washington, D.C., 1951.

[30] Levine, H. Skin friction on a strip of finite width moving parallel to its length, *J. Fluid Mech.*, 3, 147-158, 1957.

[31] Macaskill, C.C. Numerical Solution of Some Fluid-flow Problems by Boundary-Integral-Equation Techniques, Ph.D. Thesis, University of Adelaide, 1977.

[32] Macaskill, C.C. and Tuck, E.O. Evaluation of the acoustic impedance of a screen, *J. Aust. Math. Soc.*, Ser. B, 20, 46-61, 1977.

[33] MacCamy, R.C. On the heaving motion of cylinders of shallow draft, *J. Ship Res.*, 5, 34-43, 1961.

[34] Mei, C.C. Radiation and scattering of transient gravity waves by vertical plates, *Quart. J. Mech. Appl. Math.*, 19, 417-440, 1966.

[35] Mei, C.C. Numerical methods in water-wave diffraction and radiation, *Ann. Rev. Fluid Mech.*, 10, 393-416, 1978.

[36] Mittra, R. (ed). Computer Techniques for Electromagnetics, Pergamon, 1973.

[37] Monacella, V.J. On ignoring the singularity in numerical evaluation of Cauchy principal value integrals, DTNSRDC Rep . 2356, Dept. Navy, Washington D.C., 1967.

[38] Morland, L.W. Singular integral equations with logarithmic kernels, *Mathematika*, 17, 47-56, 1970.

[39] Morse, P.M. and Rubenstein, P.J. The diffraction of waves by ribbons and slots, *Phys. Rev.*, 54, 895-898, 1938.

[40] Munk, M. General theory of thin wing sections, NACA, T.R. 142, 1922.

[41] Muskhelishvili, N.E. Singular Integral Equations, Noordhoff, 1953.

[42] Oertel, R.P. The Steady Motion of a Flat Ship, Including an Investigation of Local Flow Near the Bow, Ph.D. Thesis, University of Adelaide, 1975.

[43] Parlett, B. Progress in numerical analysis, *SIAM Rev.*, 20, 443-456, 1978.

[44] Porter, D. The transmission of surface waves through a gap in a vertical barrier, *Proc. Camb. Phil. Soc.*, 71, 411-421, 1972.

[45] Prandtl, L. Aerofoil theory, I and II, *Commun. Nachr. Kgl. Gesellschaft der Wissenschaft*, Math. Phys., Gottingen, p.451, 1918. (see also NACA, T.R. 116, 1921).

[46] Richmond, J.H. Scattering by a di-electric cylinder of arbitrary cross section shape, *Trans IEEE*, AP-13, 334-341, 1965.

[47] Sayer, P. and Ursell, F. Integral equation methods for calculating the virtual mass in water of finite depth, *Proc. 2nd Int. Conf. Num. Ship Hydro.*, University of California, Berkeley, 1977.

[48] Siew, P.F. and Hurley, D.G. Long surface waves incident on a submerged horizontal plate, *J. Fluid Mech.*, 83, 141-151, 1977.

[49] Sih, G.C. (ed.). Mechanics of Fracture, Vol.1: Methods of analysis and solution of crack problems, Noordhoff, 1973.

[50] Taylor, P.J. The Effect of Beam Seas on a Stationary Ship in Shallow Water, Ph.D. Thesis, University of Adelaide, 1971.

[51] Taylor, P.J. The blockage coefficient for flow about an arbitrary body immersed in a channel, *J. Ship Res.*, 17, 97-105, 1973.

[52] Tricomi, F.G. Integral Equations, *Interscience*, 1957.

[53] Tuck, E.O. A simple Filon-trapezoidal rule, *Math. Comp.*, 21, 239-241, 1967.

[54] Tuck, E.O. Calculation of unsteady flows due to small motions of cylinders in a viscous fluid, *J. Eng. Math.*, 3, 29-44, 1969.

[55] Tuck, E.O. Ship motions in shallow water, *J. Ship Res.*, 14, 317-328, 1970.

[56] Tuck, E.O. Irrotational flow past bodies close to a plane surface, *J. Fluid Mech.*, 50, 481-491, 1971.

[57] Tuck, E.O. A theory for the design of thin heat flux meters, *J. Eng. Math.*, 6, 355-368, 1972.

[58] Tuck, E.O. Low-aspect-ratio flat-ship theory, *J. Hydronautics*, 9, 3-12, 1975.

[59] Tuck, E.O. The effect of span-wise oscillations on the thrust-generating performance of a flapping thin wing, *Symp. Swimming and Flying in Nature*, California Inst. Tech. 1974, Proceedings, Plenum Press, Vol. II, 953-974.

[60] Tuck, E.O. A note on yawed slender wings, *J. Eng. Math.*, 13, 47-62, 1979.

[61] Tuck, E.O. and Newman, J.N. Hydrodynamic interactions between ships, *10th Symp. Naval Hydro*, M.I.T., 1974. Proceedings, O.N.R. Washington D.C. 35-70.

[62] Van Dyke, M. Perturbation Methods in Fluid Mechanics, Academic Press, 1964.

[63] Wang, H.T. Comprehensive evaluation of six thin-wing lifting-surface computer programs, DTNSRDC Rep. 4333, Dept. Navy, Washington, D.C. 1974.

[64] Wood, W.W. and Head, A.K. The motion of dislocations, *Proc. Roy. Soc. Lond.*, Ser. A., 336, 191-209, 1974.

[65] Yeung, R.W. and Tan, W.T. Hydrodynamic interactions of ships with fixed obstacles, *J. Ship. Res.*, 1979.

[66] Yokobori, T. and Ichikawa, M. The interaction of parallel elastic cracks and parallel slip bands respectively, based on the concept of continuous distribution of dislocations, *Rep. Res. Inst. Strength and Fracture of Materials*, Tohoku University, 3, 1-37, 1967.

[67] Zienkiewicz, O., Kelly, D.W. and Bettes, P. The coupling of the finite element method and boundary solution procedures, *Int. J. Num. Meth. Eng.*, 11, 355-375, 1977.

A REVIEW OF NUMERICAL METHODS FOR FREDHOLM EQUATIONS OF THE SECOND KIND

Ian H. Sloan
School of Mathematics, University of New South Wales, Sydney, N.S.W.

1. Introduction

The aim of this paper is to present a rapid survey of numerical methods for linear Fredholm integral equations of the second kind - that is, for equations of the form

1.1 $$y(t) = f(t) + \int_a^b k(t,s)y(s)ds \ , \quad a \le t \le b \ ,$$

where the inhomogeneous term f and the kernel k are given. The interval $[a,b]$ may be finite or infinite. Moreover, for many important practical applications the interval $[a,b]$ should be understood to be replaced by an appropriate region of a two or three dimensional space.

We shall assume throughout that the solution of (1.1), if it exists, is unique, or correspondingly, that the homogeneous equation corresponding to (1.1) has only the trivial solution. Thus the eigenvalue problem will be outside our compass; but of course many of the methods we will be considering are also relevant to the eigenvalue case.

We shall say that an integral equation of the form (1.1) is a Fredholm equation only if it possesses the nice mathematical properties discovered (in a special case) by the Swedish mathematician Ivar Fredholm. The additional constraint to be imposed on the kernel will be described below. For the present, it might be helpful to note two commonly occurring classes of integral equations which are *not* Fredholm equations: first, the singular integral equations of Cauchy type discussed earlier in this symposium; and second, integral equations over infinite intervals with kernels of the form $k(t,s) = \kappa(t-s)$.

In the following sections a variety of methods for Fredholm equations of the second kind will be discussed. Much of the emphasis will be on practical aspects of the methods, and proofs of theoretical results will generally be omitted; but I hope that at least some readers will be encouraged to delve more deeply into the theory, and to that end I have tried to make the references as complete as possible.

Of previous reviews and books that consider equations of the second kind, the most useful are the books by Atkinson [5] and Baker [8]. (The latter contains a great deal of interesting material, but arrived on my desk too late to influence this paper.) Reviews concerned with particular methods for equations of the second kind are [7, 25, 30]. Other reviews and books containing useful material on the numerical solution of equations of the second kind are [11, 12, 18, 31, 45]. Noble [32] has given an extensive bibliography of work before 1971.

We conclude the introductory remarks with some mathematical preliminaries, treated as briefly as possible. A fuller account may be obtained, for example, from [5, Part 1].

The analysis of particular numerical methods for (1.1) is almost always expressed in the language of function spaces, for example the space $C[a,b]$ of continuous functions on a finite interval $[a,b]$, or the space $L_2(a,b)$ of square-integrable functions on a finite or infinite interval. The choice of a function space is determined partly by the particular integral equation and partly by the numerical method. Usually we require both f and y to belong to the space, and the integral operator K, defined by

1.2 $$Kg(t) = \int_a^b k(t,s)g(s)ds , \quad a \le t \le b ,$$

to be such that if g belongs to the space then so does Kg - that is, we require K to be a linear operator in the particular space. (A further condition on K will be imposed later in this section.) The integral equation (1.1) can now be written as $y = f + Ky$, or as $(I-K)y = f$, where I denotes the unit operator in the particular space.

The function spaces of interest to us are equipped with a norm, and moreover are 'complete' (meaning that every Cauchy sequence converges) - in short, they are Banach spaces. In the space $C[a,b]$ the norm is the uniform norm

1.3 $$\|g\|_\infty = \max_{a\le t\le b} |g(t)| ,$$

whereas in the space $L_2(a,b)$ the norm is

1.4 $$\|g\|_2 = \left[\int_a^b |g(t)|^2 dt\right]^{\frac{1}{2}} .$$

The norm provides a convenient tool for describing convergence: a sequence of approximate solutions y_n is said to converge to y if $\|y_n-y\| \to 0$ as $n \to \infty$, where the norm is that which is appropriate to the space under consideration.

We shall assume that the integral operator K is a compact (or completely continuous) linear operator in the particular Banach space. The mathematical definition of a compact operator (that it maps every bounded set to a set with compact closure: or that it maps every bounded sequence to a sequence with a convergent subsequence) is perhaps not very illuminating. The real point is that with this assumption the so-called Fredholm alternative holds for the integral equation (1.1). This states, in part, that if (as we are assuming in this paper) the homogeneous equation corresponding to (1.1) has no non-trivial solution belonging to a Banach space B, and if K is a compact linear operator in B, *then a (unique) solution* $y \in B$ *exists for each* $f \in B$. The Fredholm alternative therefore entirely settles the question of the existence of a solution. For a complete statement of the Fredholm alternative see [27, p.497]. Formally, the solution y can be written as

$$y = (I-K)^{-1}f .$$

A theorem of functional analysis, the Banach isomorphism theorem, then guarantees that the operator $(I-K)^{-1}$ is bounded, i.e. $\|(I-K)^{-1}\| < \infty$.

Obviously, it is important to be able to recognize compact integral operators. In $L_2(a,b)$ an easily tested sufficient condition is

$$\int_a^b \int_a^b |k(t,s)|^2 dtds < \infty . \tag{1.5}$$

In $C[a,b]$ (with $[a,b]$ finite) a sufficient condition is (see [5, p.106])

$$\sup_{t\in[a,b]} \int_a^b |k(t,s)|ds < \infty , \tag{1.6}$$

together with

$$\lim_{t\to\tau} \int_a^b |k(t,s) - k(\tau,s)|ds = 0 , \quad a \le \tau \le b . \tag{1.7}$$

The latter property is not always easy to verify directly, but can often be easily inferred by using techniques described in [22].

If the interval is finite, it follows from the above that a kernel k that is continuous on $[a,b] \times [a,b]$ yields a compact integral operator in both $C[a,b]$ and $L_2(a,b)$. A kernel of the weakly singular form $k(t,s) = |t-s|^{-\alpha}m(t,s)$, where m is continuous and non-zero, satisfies the condition (1.5) only if $\alpha < \frac{1}{2}$, but can be shown (see [22]) to satisfy (1.6) and (1.7) (and hence to be compact in $C[a,b]$) for all $\alpha < 1$.

2. The Nyström Method

Suppose that a and b are finite, and that the inhomogeneous term f and the kernel k are continuous (and preferably smoother). Then the simplest way to tackle the integral equation (1.1) is to approximate the integral by one of the standard quadrature rules (e.g. the Simpson or Gauss rules). We write the quadrature rule as

2.1 $$\int_a^b g(s)ds \approx \sum_{i=1}^{n} w_{ni}g(t_{ni}) ,$$

where $\{t_{ni}\}_{i=1}^n$ and $\{w_{ni}\}_{i=1}^n$ are the points and weights for the particular n-point quadrature rule on the interval $[a,b]$.

If y_n denotes the approximate solution of (1.1) obtained by replacing the integral in (1.1) by the quadrature rule (2.1), then y_n satisfies

2.2 $$y_n(t) = f(t) + \sum_{i=1}^{n} w_{ni}k(t,t_{ni})y_n(t_{ni}) .$$

A set of n linear equations for the n unknowns $y_n(t_{n1}),\ldots,y_n(t_{nn})$ may then be obtained by setting $t = t_{nj}$, $j = 1,\ldots,n$, to give

2.3 $$\sum_{i=1}^{n} [\delta_{ji}-w_{ni}k(t_{nj},t_{ni})]y_n(t_{ni}) = f(t_{nj}) , \quad j = 1,\ldots,n ,$$

where δ_{ji} is the Kronecker delta.

This set of linear equations, together with (2.2) (which we may think of as a natural interpolation formula), constitute the Nyström method [33].

A very satisfactory convergence theory exists for the Nyström method on a finite interval, and is set out, for example, in [5, pp.88-104]. The only assumption required in the theory, beyond the continuity of f and k , is that the integration rule (2.1) should converge to the exact integral as $n \to \infty$ for *all* continuous functions g . (That property certainly holds for the Simpson and Gauss rules, and also for most of the other familiar integration rules, but does not hold for the nth-order Newton-Cotes rule.) The analysis, carried out in the space $C[a,b]$, then shows that y_n converges to y as $n \to \infty$; or in other words $y_n(t)$ converges uniformly to $y(t)$.

Of course we are very interested also in the *rate* of convergence. Let us define a linear operator K_n in the space $C[a,b]$ by

2.4 $$K_n g(t) = \sum_{i=1}^{n} w_{ni}k(t,t_{ni})g(t_{ni}) , \quad g \in C[a,b] ,$$

so that (2.2) can be written $y_n = f + K_n y_n$, and K_n becomes our numerical quadrature approximation to the integral operator K defined by (1.2). Since the Nyström method in effect replaces K by K_n in the exact equation $y = f + Ky$, we presumably expect the rate of convergence to be limited by the rate at which $K_n y$ approaches Ky , where y is the exact solution. That rate of convergence is actually achieved, in the sense that the theory yields (see below)

$$\|y_n - y\| \le M\|K_n y - Ky\| ,$$

where M is a constant. (The norm in this section is the uniform norm defined by (1.3).) Thus the integration rule should be chosen so that it integrates efficiently $k(t,s)y(s)$ as a function of s , for each fixed $t \in [a,b]$.

Though we can only hint at the theory here, it should be mentioned that the most elegant approach to the theory (and it is one that extends also to most other numerical methods for equations of the second kind), is Anselone's collectively compact operator approximation theory [1]. For that theory to be applicable, one has to show that $\|K_n g - Kg\| \to 0$ as $n \to \infty$ for all continuous functions g , and that the sequence $\{K_n\}$ has a property known as collective compactness [1 p.4, 5 p.96]; both properties hold for the quadrature operator (2.4), provided the rule (2.1) converges to the exact integral for all $g \in C[a,b]$. It then follows from the theory that $(I-K_n)^{-1}$ exists for all n sufficiently large, and that there exists $M > 0$ and an integer n_0 such that for all $n \ge n_0$ we have

$$\|(I-K_n)^{-1}\| \le M .$$

The rest of the error analysis is then quite straightforward: for $n \ge n_0$ we have

2.5 $$y_n = (I-K_n)^{-1} f$$

as well as

2.6 $$y = (I-K)^{-1} f ,$$

from which it follows that

$$y_n - y = (I-K_n)^{-1}(K_n y - Ky) ,$$

and hence

$$\|y_n - y\| \le \|(I-K_n)^{-1}\| \, \|K_n y - Ky\|$$

$$\le M\|K_n y - Ky\| .$$

The Nyström method is easily programmed for the computer, but obviously can become very time consuming if n is large. However, the time required can be reduced to manageable proportions by using one or other of the available iterative techniques [4, 10, 29].

In this connection, mention should be made of a careful program for the Nyström method developed by Atkinson [6]. The program uses either Simpson's rule or the Gauss rule, and, if appropriate, solves the resulting system of linear equations by means of an iterative method due to Brakhage [10]. In brief, the iterative method is as follows: First, choose a suitable integer n of moderate size, for which the equations (2.3) can be solved directly. For $m > n$ we write

$$\begin{aligned} y_m &= f + K_m y_m \\ &= f + (K_m - K_n) y_m + K_n y_m \\ &= f + (K_m - K_n)(f + K_m y_m) + K_n y_m , \end{aligned}$$

and this equation is solved iteratively after taking the last term of the right-hand side over to the left; i.e. given $y_m^{(0)}$, one constructs $y_m^{(1)}, y_m^{(2)}, \ldots,$ by

2.7 $$(I - K_n) y_m^{(k+1)} = f + (K_m - K_n) f + (K_m - K_n) K_m y_m^{(k)} .$$

Convergence of the iterative scheme is guaranteed, if n is sufficiently large. Practical details of the iterative scheme are given in [4, 6].

What integration rules should be used for the Nyström method? For finite intervals perhaps the most popular choices are Simpson's rule and the Gauss rule. However, in place of the latter, I recommend the Clenshaw-Curtis rule [15]. (The use of the Clenshaw-Curtis rule has also been suggested by El-Gendi [19].) Practical experience suggests that the Clenshaw-Curtis rule is nearly as accurate as the Gauss rule, yet the points (if the interval is mapped onto $[-1,1]$) are given by the remarkably simple formula

2.8 $$t_{nj} = \cos \frac{(j-1)\pi}{n-1} , \quad j = 1, \ldots, n ,$$

and the formula for the weights (see [26]) is merely a simple finite sum. A further advantage of the Clenshaw-Curtis rule is that useful error estimates can easily be computed (see [15, 43]); but apparently such estimates have not yet been applied to integral equations.

In all of the foregoing, the Nyström method has been restricted to the case of finite intervals. Yet many of the integral equations that occur in practice have infinite integration regions. In many such cases the Gauss-Hermite or Gauss-Laguerre rules would seem very appropriate, and they are often used in practice, but as far as I am aware there exists no error analysis or convergence theory that would justify their use. That is a challenge for the theorists in the audience.

3. Product Integration

Suppose the kernel takes the form

3.1 $$k(t,s) = h(t,s)m(t,s) ,$$

where the second factor $m(t,s)$ is a smooth function of s for each t in $[a,b]$, whereas $h(t,s)$ is singular (or if not singular, at least not very smooth). Then the Nyström method will generally be unsatisfactory, either not converging at all, or converging only slowly. On the other hand, one of the variants of the product-integration method, pioneered by Young [46] and Atkinson [2,3], may be very appropriate. This is particularly so if the singular factor $h(t,s)$ can be chosen to be a function of a standard type, such as $\ell n|t-s|$ or $|t-s|^\alpha$, $\alpha > -1$, and this can often be achieved by taking advantage of the freedom that exists in choosing the smooth factor $m(t,s)$. If a suitable factorization in (3.1) can be achieved, then the product-integration approach, now to be described, is likely to be most successful, combining good accuracy with practical efficiency. (There is a slight catch, however, which we shall come to at the end of this section.)

The product-integration method is similar to the Nyström method, in that the integral term in (1.1) is approximated by a quadrature rule. The difference here is that the quadrature rule is of the so-called product-integration form: it is a rule specially designed to accommodate the singular factor $h(t,s)$, hence that factor does not appear explicitly in the quadrature sum, but rather is incorporated into the weights.

We recall that most ordinary integration rules are based on approximating the integrand by an interpolating function of some kind (for example piecewise-linear for the trapezoidal rule, polynomials for the Gauss or Newton-Cotes rules) and then carrying out the integration exactly. A product-integration rule is obtained by approximating just the smooth part of the integrand by an interpolating function, and again integrating exactly.

More precisely, given $n > 0$ we select a set of n quadrature points $t_{n1},\ldots,t_{nn}$ in the interval $[a,b]$, and a corresponding n-dimensional space of interpolating functions, which we will assume has a basis $\{\ell_{n1},\ldots,\ell_{nn}\}$ satisfying

3.2 $$\ell_{ni}(t_{nj}) = \delta_{ji} .$$

(For example, for the ordinary Gauss rule the points t_{ni} are the zeros of the Legendre polynomial $P_n((2t-a-b)/(b-a))$, the n-dimensional space is the space of all polynomials of degree $\le n-1$, and the functions ℓ_{ni} are the fundamental polynomials of Lagrange interpolation for the points t_{ni}. For the trapezoidal rule the points t_{ni} are equally spaced, and the space is a space of continuous piecewise-linear functions, with possible discontinuities in the first derivative only at the points t_{ni}.) Given a function $g \in C[a,b]$, we denote by $p_n g$ the unique function in the n-dimensional space that coincides with g at $t_{n1},\ldots,t_{nn}$. It is easily seen that

$$p_n g(s) = \sum_{i=1}^{n} \ell_{ni}(s) g(t_{ni}) \ ; \tag{3.3}$$

for it then follows immediately from (3.2) that

$$p_n g(t_{nj}) = g(t_{nj}) \ , \quad j = 1,\ldots,n \ .$$

Since we are here seeking an approximate solution for

$$y(t) = f(t) + \int_a^b h(t,s) m(t,s) y(s) ds \ , \quad a \le t \le b \ ,$$

we approximate just the smooth (with respect to s) factor $m(t,s)y(s)$ by the interpolatory approximation (3.3). Thus the approximate equation is

$$\begin{aligned} y_n(t) &= f(t) + \int_a^b h(t,s) \sum_{i=1}^{n} \ell_{ni}(s) m(t,t_{ni}) y_n(t_{ni}) ds \\ &= f(t) + \sum_{i=1}^{n} w_{ni}(t) m(t,t_{ni}) y_n(t_{ni}) \ , \end{aligned} \tag{3.4}$$

where

$$w_{ni}(t) = \int_a^b h(t,s) \ell_{ni}(s) ds \ . \tag{3.5}$$

Equation (3.4) is analogous to the Nyström equation (2.2). In just the same way we obtain from (3.4) a set of linear equations for the unknowns $y_n(t_{n1}),\ldots,y_n(t_{nn})$,

$$\sum_{i=1}^{n} [\delta_{ji} - w_{ni}(t_{nj}) m(t_{nj},t_{ni})] y_n(t_{ni}) = f(t_{nj}) \ , \quad j = 1,\ldots,n \ , \tag{3.6}$$

and then use (3.4) as a natural interpolation formula for values of $y_n(t)$ between the quadrature point.

The only substantially new element from a practical point of view is the necessity for calculating the weights $w_{ni}(t)$, defined by (3.5). Atkinson [2, 3] has discussed this question in detail for the particular cases of piecewise-linear and piecewise-quadratic interpolating functions (the method being referred to in these cases as the product-trapezoidal and product-Simpson method respectively). We refer the reader to the original papers for details. (In brief, Atkinson shows that for a kernel such as $h(t,s) = \ell n|t-s|$, the calculation of the $n \times n$ matrix $\{w_{ni}(t_{nj})\}$ which is required in (3.6) can be accomplished quite economically for the piecewise-linear and piecewise-quadratic cases, because the total number of independent integrals that must be evaluated to handle all values of n up to say, $n = 50$ is quite small. It may be remarked, however, that Atkinson's discussion does not consider the calculation of the weights $w_{ni}(t)$ for arbitrary values of t , and hence does not allow for the practical use of (3.4) as an interpolation formula.)

Atkinson [2, 3] also gives a very complete convergence proof and error analysis for the product-trapezoidal and product-Simpson methods. In particular, he shows that provided f and m are continuous, and that h satisfies (1.6) and (1.7), then y_n exists for all n sufficiently large, and converges uniformly to y. The argument is again based on the Anselone collectively compact operator approximation theory, so that all of the remarks made about the error analysis in the previous section apply again here, provided we re-define K_n to be

$$K_n g(t) = \sum_{i=1}^{n} w_{ni}(t)m(t,t_{ni})g(t_{ni}) , \quad g \in C[a,b] .$$

In Atkinson's discussion of product-integration methods, the interpolating functions are restricted to be piecewise polynomials. It turns out, however, that there is also something to be said for choosing the interpolating functions to be polynomials, and I shall conclude this section with a brief mention of some recent work of my own in this area.

Though the polynomial case requires rather different theoretical techniques from the piecewise-polynomial cases considered above, a satisfactory convergence theory can in fact be established, under quite mild conditions on the function $h(t,s)$ in (3.1), provided the quadrature points t_{ni} are carefully chosen. One suitable choice, if the interval $[a,b]$ is taken for convenience to be $[-1,1]$, is

3.7 $$t_{ni} = \cos\frac{(2i-1)\pi}{2n} , \quad i = 1,\dots,n ,$$

the points so defined being the zeros of the Chebyshev polynomial of the first kind $T_n(x)$. With that choice the underlying product-integration rule is one that has been discussed in detail in [39], and for suitable functions $h(t,s)$ the weights $w_{ni}(t)$ can be calculated (for any value of t, whether a quadrature point or not) by the techniques developed in that paper.

To illustrate the polynomial product-integration method, the method has been applied to the integral equation

3.8 $$y(t) = f(t) + \int_{-1}^{1} |t-s|^{-\frac{1}{2}}(t^2+s^2)y(s)ds , \quad -1 \le t \le 1 ,$$

with f chosen so as to make the exact solution $y(s) = (1-s)^{\frac{1}{2}}$. (This solution, though contrived, has realistically nasty behaviour for at least one end of the interval, and therefore would seem to be a fair test.) The singular factor h in (3.1) was chosen to be

$$h(t,s) = |t-s|^{-\frac{1}{2}} ,$$

so that

$$m(t,s) = t^2 + s^2 ,$$

which is indeed smooth. The results obtained, shown in Table 1, seem to be quite satisfactory for an example of this kind.

Table 1

n	$\|y-y_n\|_\infty$
8	0.54
16	0.0077
24	0.0026
32	0.0012
40	0.00072

A full account of the theory and application of the polynomial product-integration method is in preparation.

The example (3.8) draws attention to a feature of so-called weakly singular integral equations that is often overlooked in discussions of numerical methods, namely that a singularity in the kernel tends to produce non-analytic behaviour in the *solution* $y(t)$ at the end-points of the interval. In contrast, many of the examples that have been used in the past to test product-integration methods have solutions that are analytic over the whole interval. Such examples tend to give much too favourable a view of the prospects for product-integration methods in practice.

4. Collocation Method and Variants

The collocation method is at first sight rather different from the foregoing methods, in that it is an expansion method: for each n one selects a basis set $u_{n1},\ldots,u_{nn}$ of real-valued functions defined on $[a,b]$, and approximates y by a linear combination

4.1 $$y_n = \sum_{i=1}^{n} a_{ni}u_{ni} .$$

The coefficients a_{ni} are then fixed by 'collocating' at n selected points $t_{n1},\ldots,t_{nn}$ in the interval, that is by requiring

4.2 $$y_n(t_{nj}) = f(t_{nj}) + \int_a^b k(t_{nj},s)y_n(s)ds , \quad j = 1,\ldots,n .$$

We immediately obtain a set of n linear equations for the coefficients $a_{n1},\ldots,a_{nn}$,

4.3 $$\sum_{i=1}^{n} [u_{ni}(t_{nj}) - \int_a^b k(t_{nj},s)u_{ni}(s)ds]a_{ni} = f(t_{nj}) , \quad i = 1,\ldots,n .$$

In practice, usually the most difficult part of the collocation method is the evaluation of the integrals on the left-nand side of (4.3).

A variant of the above, suggested by the form of the exact equation $y = f + Ky$, is the approximation of y by

$$Y_n = f + \sum_{i=1}^{n} b_{ni} u_{ni} , \tag{4.4}$$

instead of (4.1). The collocation procedure then leads to a set of linear equations with the same left-hand side matrix as (4.3), but with the right-hand side replaced by $\int_a^b k(t_{nj},s)f(s)ds$. The two methods are easily seen to be equivalent if f is a linear combination of $u_{n1},\ldots,u_{nn}$. In the following we consider only the first form of the collocation method.

A more interesting variant is obtained by substituting the collocation solution (4.1) into the right-hand side of (1.1), to obtain a new approximation y'_n , which we may refer to as the iterated collocation method, and which is given by

$$y'_n(t) = f(t) + \int_a^b k(t,s)y_n(s)ds \tag{4.5}$$

$$= f(t) + \sum_{i=1}^{n} a_{ni} \int_a^b k(t,s)u_{ni}(s)ds , \tag{4.6}$$

where the coefficients a_{ni} are given by (4.3).

The most important property of y'_n follows immediately from (4.2) and (4.5), namely that y'_n coincides with y_n at the collocation points (or nodes) t_{nj} , that is

$$y'_n(t_{nj}) = y_n(t_{nj}) , \qquad j = 1,\ldots,n . \tag{4.7}$$

Thus we may regard y'_n and y_n merely as different interpolation devices for interpolating between the approximate values of y obtained at the nodes. We are therefore free, if we are only interested in the behaviour at the nodes, to analyse the convergence behaviour of the collocation method in terms of whichever of y_n and y'_n gives the most useful results. The iterated collocation solution may therefore be useful in the analysis, even if never calculated explicitly.

The iterated collocation method described above is in fact just a special case of the product-integration method described in the previous section: It corresponds to the case in which $h(t,s) = k(t,s)$ and $m(t,s) = 1$. The correspondence is most easily demonstrated if the basis set $u_{n1},\ldots,u_{nn}$ happens to satisfy $u_{ni}(t_{nj}) = \delta_{ij}$; for it then follows from (4.1) and (4.7) that

4.8 $$a_{ni} = y_n(t_{ni}) = y_n'(t_{ni}) ,$$

and with this substitution (4.6) becomes identical with (3.4), if we identify ℓ_{ni} with u_{ni}. That simple argument is invalid if $u_{ni}(t_{nj}) \neq \delta_{ij}$, but since the collocation approximation y_n can easily be seen to be invariant (apart from rounding errors) under a change of basis in the space spanned by $u_{n1},\ldots,u_{nn}$, the above conclusion holds as long as there exists a basis $\ell_{n1},\ldots,\ell_{nn}$ with the property $\ell_{ni}(t_{nj}) = \delta_{ij}$. A necessary and sufficient condition for such a basis to exist is that the $n \times n$ matrix $\{u_{ni}(t_{nj})\}$ be non-singular; a condition that is certainly essential if the collocation method is to have any reasonable hope of success.

The principal decision that has to be made in using the collocation method concerns the choice of the basis set u_{n1},u_{nn}, and the closely related question of the choice of the nodes $t_{n1},\ldots,t_{nn}$. The breadth of the range of possible choices for the basis set gives great richness and flexibility to the collocation method, but naturally somewhat complicates the theoretical analysis.

Broadly, there are two main classes of functions from which the basis set may be chosen: first, piecewise-polynomials (e.g. piecewise-linear functions, splines etc.) and second, functions defined by a single expression over the whole interval (e.g. polynomials, trigonometric functions).

The general theoretical situation for the piecewise-polynomial case is that the convergence of the collocation approximation y_n to the exact solution y can usually be established without difficulty. It is of course required that the nodes $\{t_{nj}\}$ be sensibly chosen; in fact, they should be chosen so that for every $g \in C[a,b]$, the unique linear combination of $u_{n1},\ldots,u_{nn}$ that coincides with g at $t_{n1},\ldots,t_{nn}$ should converge uniformly to g as $n \to \infty$. That condition can usually be achieved without difficulty for the piecewise-polynomial case. (See below for the piecewise-constant case.) Standard arguments (see for example [25]) can then be used to prove the uniform convergence of y_n to y, provided merely that f is continuous and k satisfies (1.6) and (1.7).

Detailed treatments of the collocation method with piecewise-polynomial basis functions have been given by Ikebe [25] (piecewise-constant and piecewise-linear), Phillips [34] (cubic splines), Houstis and Papatheodorou [24] (piecewise cubics), and Prenter [36].

For the case of polynomials, on the other hand, the theoretical situation is more complicated, as the theory of [25] does not apply. The point is that it is *impossible* (see [16, p.78-79]) to choose the nodes $\{t_{nj}\}$ so that the unique polynomial of degree $< n$ that coincides with g at the nodes $t_{n1},\ldots,t_{nn}$ converges uniformly to g for all continuous functions g. Nevertheless, it is well known that some choices of nodes have very much better interpolation properties than others, and that among the best are the Chebyshev nodes (3.7) and the Clenshaw-Curtis nodes (2.8). (We here assume for convenience that the interval $[a,b]$ is $[-1,1]$.) The equally spaced points, on the other hand, are known to have rather bad interpolation properties (see [16, p.78]), and should therefore not be used when working with polynomials.

If the nodes in the polynomial case are taken to be the Chebyshev points, then the theory of Phillips [34] can be used to prove the uniform convergence of y_n to y, provided firstly that f is such that the polynomial of lowest degree that interpolates f at the Chebyshev points converges uniformly to f, and secondly that k satisfies (1.6) and (1.7), and also an additional rather complicated condition which I shall not describe, but which though not satisfied even for all continuous kernels, is in fact satisfied for most of the continuous and even weakly singular kernels that arise in practice.

An alternative and perhaps cleaner treatment of the polynomial case for the Chebyshev nodes, based on the iterated collocation approximation y_n' rather than on y_n itself, has been given by Sloan and Burn [40]. They show that y_n' converges uniformly to y if f is merely continuous and if k satisfies

$$\sup_{t\in[-1,1]} \int_{-1}^{1} |k(t,s)|^p ds < \infty$$

and

$$\lim_{t\to\infty} \int_{-1}^{1} |k(t,s) - k(\tau,s)|^p ds = 0 , \quad -1 \le \tau \le 1 ,$$

for some $p > 1$; conditions that are only fractionally stronger than (1.6) and (1.7).

Practical aspects of the collocation method with polynomials are also discussed in [40], and a computational criterion is given there for deciding which of y_n and y_n' is likely to be the more accurate in a given practical situation.

When working with polynomials in Chebyshev nodes, a convenient set of basis functions is

$$u_{ni}(t) = T_{i-1}(t) , \quad i = 1,\ldots,n ,$$

where T_j is the Chebyshev polynomial of the first kind defined by $T_j(\cos\theta) = \cos j\theta$. The first advantage of this choice of basis is that it leads to a well conditioned matrix on the left-hand side of (4.3). The second is that for many of the common kernels, including even the weakly singular kernel $k(t,s) = |t-s|^\alpha$, $\alpha > -1$, a convenient recursive technique for calculating the integrals

4.9 $$\int_{-1}^{1} k(t,s)T_j(s)ds , \quad j \ge 0 ,$$

which are needed on the left-hand side of (4.3) and also possibly in (4.6), has been developed by Piessens and Branders [35]. These authors give recurrence relations for a number of important kernels, together with recommendations for the use of the recurrence relations, based on a careful analysis of their stability properties.

(It should be remarked, however, that the full method described by Piessens and Branders in [35] is a collocation method using polynomials, but with equally spaced points. That method should not be used by those who like to have a theoretical foundation for their numerical calculations, as there appears to be no theory that covers this case.)

If the kernel k is smooth enough to allow the integrals in (4.9) to be evaluated numerically, then Elliott's variant [20] of the collocation method with polynomials may be appropriate. In that method the nodes $\{t_{nj}\}$ are the Clenshaw-Curtis points (2.8), and the integrals (4.9) are evaluated approximately by what is in effect the Clenshaw-Curtis quadrature rule [15] of order m, where m need not be the same as n. A detailed algorithm for this variant, with the additional refinement of allowing for a discontinuity in $k(t,s)$ along the line $t = s$, has been given in [2].

If we take the basis functions to be trigonometric functions instead of polynomials, for example (for the interval [0,1])

$$u_{ni}(t) = \sin i\pi t \ , \quad i = 1,\dots,n \ ,$$

then the situation seems to be broadly the same as for the polynomial case, except that the nodes should presumably be taken to be equally spaced, say

$$t_{ni} = \frac{i}{n+1} \ , \quad i = 1,\dots,n \ ,$$

the point being that for trigonometric interpolation equally spaced points are the safest. However, I know of no explicit treatment of the collocation method with trigonometric functions.

To conclude this section, let us return to look more closely at the simplest piecewise-polynomial case, that of piecewise-constant basis functions. Given $n > 0$, we first have to define a suitable partition of the interval $[a,b]$,

$$a = s_0 < s_1 < \dots < s_{n-1} < s_n = b$$

(where $s_i = s_{ni}$, with the label n omitted to simplify the notation). Then we approximate y by a function y_n that is constant within each sub-interval $[s_{i-1}, s_i]$, with the value of y_n within each sub-interval determined by the collocation procedure (4.2). It turns out, as we shall see, that with this approximation it pays to take special care in choosing the collocation points $\{t_{nj}\}$.

A convenient basis set for the piecewise-constant case is

$$u_{ni}(t) = \begin{cases} 1 & \text{if} \quad s_{i-1} < t \le s_i \ , \\ 0 & \text{otherwise} \ , \end{cases}$$

for $i = 2,\ldots,n$, and

$$u_{n1}(t) = \begin{cases} 1 & \text{if} \quad s_0 \le t \le s_1 , \\ 0 & \text{otherwise} . \end{cases}$$

The integrals on the left-hand side of (4.3) then reduce to integrals over a single sub-interval, since

$$\int_a^b k(t,s)u_{ni}(s)ds = \int_{s_{i-1}}^{s_i} k(t,s)ds .$$

The piecewise-constant collocation method might be thought too crude to be of any practical interest. Yet its extension to two and three dimensions is very widely used, perhaps more widely than any other method when it comes to really hard practical problems. (For higher dimensions the sub-interval $[s_{i-1},s_i]$ should be replaced by a sub-region S_{ni} .) Thus anything we can do to improve the efficiency of the piecewise-constant method would seem worthwhile.

We now ask: how should the nodes $\{t_{nj}\}$ be chosen? As far as the convergence theory of Ikebe [25] is concerned, the only restrictions are that there should be exactly one node within each sub-interval, and that the length of the largest sub-interval should converge to zero, i.e.

$$\lim_{n\to\infty} \max_{1\le i\le n} |s_i - s_{i-1}| = 0 .$$

For it is then easily verified that for any given $g \in C[a,b]$, the linear combination of $u_{n1},\ldots,u_{nn}$ that interpolates g at $t_{n1},\ldots,t_{nn}$ converges uniformly to g . Ikebe's theory [25] can then be used to prove convergence of y_n to y , provided f is continuous and k satisfies (1.6) and (1.7). For the usual case in which the sub-intervals are of equal length, the *rate* of convergence is given by

$$\|y_n - y\|_\infty = O(\tfrac{1}{n}) ,$$

which is obviously the fastest possible rate of convergence for a sequence of piecewise-constant approximations, unless y itself is actually a constant.

On the other hand, by looking at the question from the point of view of y_n' rather than y_n itself, it can be shown (see [42, Section 6]) that a faster rate of convergence at the nodes can often be achieved by a more careful choice of nodes. Specifically, if each node is chosen at the *mid-point* of a sub-interval, and if k and y satisfy certain smoothness conditions and the sub-intervals are of equal length, then

$$\|y_n'-y\|_\infty = O\left(\frac{1}{n^2}\right) .$$

It then follows from (4.7) that the collocation solution $y_n(t)$ must have a faster rate of convergence at the nodes than it does elsewhere, a property sometimes referred to in other contexts as 'superconvergence' at the nodes.

The faster rate of convergence may be understood in an informal way by noting that in this case (4.6) may be written, using (4.8), as

4.10
$$y_n'(t) = f(t) + \sum_{i=1}^{n} \int_{s_{i-1}}^{s_i} k(t,s)ds\; y_n'(t_{ni})$$
$$\equiv f(t) + K_n y_n'(t) .$$

We therefore expect the rate of convergence to be governed by that of

$$K_n y(t) - Ky(t) = \sum_{i=1}^{n} \int_{s_{i-1}}^{s_i} k(t,s)[y(t_{ni}) - y(s)]ds .$$

If $k(t,s)$ and $y(s)$ have Taylor series expansions about $s = t_{ni}$, then the leading term in the error expansion is

$$\sum_{i=1}^{n} k(t,t_{ni}) \left(\frac{dy}{dt}\right)(t_{ni}) \int_{s_{i-1}}^{s_i} (t_{ni}-s)ds .$$

This vanishes if t_{ni} is chosen at the mid-point of $[s_{i-1},s_i]$. (The situation is analogous to that for the mid-point rule in ordinary numerical quadrature.)

The simplest way to establish rigorously the higher-order rate of convergence is to view (4.10) as an equation in the space $C[a,b]$, and apply to it the Anselone collectively compact operator approximation theory [1], under the assumption that f is continuous and k satisfies (1.6) and (1.7). It then follows, as in Section 2, that the rate of convergence is that of $\|K_n y-Ky\|_\infty$, so that the higher-order rate of convergence is easily obtained by imposing suitable smoothness conditions on k and y (see [42]).

In higher dimensions a similar result (due to I.G. Graham) also holds: superconvergence at the nodes can be attained by choosing the nodes to be at the *centroids* of the subregion S_{ni}. Strictly, the result depends on both the kernel k and the solution y having a certain degree of smoothness. However, the recommendation, that each node should be chosen at the centroid of a sub-region, seems likely to be good advice even if the smoothness conditions are not satisfied.

5. Bateman Method

The Bateman method [9] is little discussed by numerical analysts, perhaps because no satisfactory convergence theory exists. (The theory of Thompson [44] can be used to prove convergence only in very special cases.) The following brief description is slightly unconventional, in that the Bateman method is presented here as a special case of the collocation method, rather than in its original formulation as a degenerate-kernel method; but the end result is the same.

Suppose k is a smooth kernel. If we wish to use the collocation method, but have an open mind about the choice of basis functions $u_{n1},\dots,u_{nn}$, one possible choice, which is in some sense natural to the problem, is

5.1 $$u_{ni}(t) = k(t,t_{ni}), \quad i = 1,\dots,n,$$

where $\{t_{ni}\}$ is a suitable set of points in $[a,b]$ satisfying

5.2 $$a \le t_{n1} < \dots < t_{nn} \le b.$$

We assume that these points are chosen so that the set $\{u_{n1},\dots,u_{nn}\}$ so obtained is linearly independent.

The Bateman method is obtained by using the basis functions (5.1) in the variant (4.4) of the collocation method, so that we have as our approximate solution

5.3 $$Y_n(t) = f(t) + \sum_{i=1}^{n} b_{ni}k(t,t_{ni}).$$

As collocation points we choose, for simplicity, the points already used in the definition (5.1). Then the equations that determine the coefficients in (5.3) are

5.4 $$\sum_{i=1}^{n} [k(t_{nj},t_{ni}) - \int_a^b k(t_{nj},s)k(s,t_{ni})ds]b_{ni} = \int_a^b k(t_{nj},s)f(s)ds.$$

We observe that the Bateman solution (5.3) has the same functional form as the Nyström solution (2.2), but with different coefficients. The matrix elements in (5.4) are more difficult to evaluate than those in the Nyström method, but in many cases it will be possible to evaluate the integral

5.5 $$\int_a^b k(t,s)k(s,r)ds$$

analytically, and in such cases the practical differences may not be great. The real difference is in the absence of a theory, so that one has no guidance in choosing the points $\{t_{nj}\}$, and no *a priori* knowledge in most cases as to whether or not the method will converge. A further disadvantage is that the matrix on the left-hand side of (5.4) will usually be not very well conditioned.

But in spite of these disadvantages, I think there might yet turn out to be a role for the Bateman method, when it is better understood. One point in its favour is that the choice of the points $\{t_{nj}\}$ is not tied to any particular quadrature rule. Thus it can be used, for instance, if f is known only at irregularly spaced points.

6. Galerkin Method and Variants

The Galerkin method, like the collocation method, is an expansion method, in which the exact solution y is approximated by

6.1 $$y_n = \sum_{i=1}^{n} a_{ni}u_{ni} .$$

The difference lies in the way the coefficients a_{ni} are determined. Let us define an inner product in our function space by

6.2 $$(g,h) = \int_a^b \overline{g(s)}h(s)\omega(s)ds ,$$

where ω is a suitable non-negative weight function, which we will take to be 1 unless stated otherwise. Then the coefficients in the Galerkin method are determined by

6.3 $$(u_{nj},y_n) = (u_{nj},f) + (u_{nj},Ky_n) , \quad j = 1,\ldots,n ,$$

giving the linear equations

6.4 $$\sum_{i=1}^{n} [(u_{nj},u_{ni}) - (u_{nj},Ku_{ni})]a_{ni} = (u_{nj},f) , \quad j = 1,\ldots,n .$$

The Galerkin method generally requires more effort than the collocation method, because of the occurrence of the double integrals

6.5 $$(u_{nj},Ku_{ni}) = \int_a^b \int_a^b \overline{u_{nj}(t)}k(t,s)u_{ni}(s)dtds$$

on the left-hand side of (6.4). Often these integrals will have to be evaluated numerically. If k is smooth, then there is at least one situation in which the calculation of the approximate integrals can be made relatively efficient: if the interval is $[-1,1]$ and the basis functions are Chebyshev polynomials, and if $\omega(s)$ is chosen to be $(1-s^2)^{-\frac{1}{2}}$, then the fast Fourier transform technique may be used, in the way described by Delves [17].

The natural setting for the theoretical analysis of the Galerkin method is the space $L_2(a,b)$ of square-integrable functions on the interval (a,b), which is a Hilbert space with the inner product (6.2) and the norm

$$\|g\|_2 = (g,g) \quad , \qquad g \in L_2(a,b) \ .$$

Given $g \in L_2(a,b)$, let us denote by $P_n g$ the orthogonal projection of g onto the subspace spanned by $u_{n1},\ldots,u_{nn}$. We shall assume that the set $\{u_{ni}\}$ is chosen in such a way that

$$\lim_{n\to\infty} \|P_n g - g\|_2 = 0$$

for all $g \in L_2(a,b)$. Then, provided $f \in L_2(a,b)$ and K is a compact operator in $L_2(a,b)$, it is known (see [31 p.293]) that

$$\|P_n y - y\|_2 \le \|y_n - y\|_2 \le (1+\delta_n)\|P_n y - y\|_2 \ ,$$

where $\delta_n \to 0$ as $n \to \infty$. Thus the Galerkin method not only converges: its error approaches asymptotically the least possible error for an approximation of the form (6.1).

The real advantage of the Galerkin method appears when we consider its iterative variant. This is defined in the same way as the iterative variant of the collocation method: if y_n denotes the Galerkin approximation, then the iterated Galerkin approximation is

6.6
$$y_n' = f + Ky_n$$
$$= f + \sum_{i=1}^{n} a_{ni} K u_{ni} \ .$$

It may be shown quite generally (see Sloan [37]) that y_n' converges to y faster than y_n does: specifically, that

6.7
$$\|y_n' - y\|_2 \le \beta_n \|y_n - y\|_2 \ ,$$

where $\beta_n \to 0$ as $n \to \infty$. The proof is omitted, though it is in fact quite simple.

For the particular case of splines, the improvement in the rate of convergence obtained by replacing y_n by y_n' has been quantified by Chandler [13]: for splines of degree r he shows, assuming f and k to be sufficiently smooth, that

$$\|y_n - y\|_2 = O\left(h^{r+1}\right) \ ,$$

whereas

$$\|y_n'-y\|_2 = O(h^{2r+2}) .$$

Even more importantly, he shows that the order of convergence may be preserved even if the integrals in (6.5) and (6.6) are evaluated numerically, provided the quadrature rules are properly chosen. But I shall leave these matters for Graeme Chandler to explain for himself.

Another variant of the Galerkin method, known in the Russian literature as the Petrov-Galerkin method, is obtained by replacing (6.3) by

$$(v_{nj},y_n) = (v_{nj},f) + (v_{nj},Ky_n) , \quad j = 1,\ldots,n ,$$

and hence replacing (6.4) by

6.8 $$\sum_{i=1}^{n} [(v_{nj},u_{ni}) - (v_{nj},Ku_{ni})]a_{ni} = (v_{nj},f) , \quad j = 1,\ldots,n ,$$

where $v_{n1},\ldots,v_{nn}$ is a second set of linearly independent functions in $L_2(a,b)$. Of course there exists an iterative variant of the Petrov-Galerkin method, defined by

$$y_n' = f + Ky_n ,$$

where y_n now denotes the Petrov-Galerkin approximation. The iterated Petrov-Galerkin approximation is equivalent to a class of degenerate kernel methods proposed by Sloan, Burn and Datyner [41].

Given the functions $\{u_{ni}\}$, the methods described in the previous paragraph give a large, indeed almost embarrassing, amount of freedom in the choice of the functions $\{v_{ni}\}$. The paper [41] considers in detail two particular choices for the set $\{v_{ni}\}$. The first choice is $v_{ni} = u_{ni}$, in which case the degenerate-kernel method becomes equivalent to the iterated Galerkin method discussed above. The second choice is $v_{ni} = K^*u_{ni}$, where K^* is the adjoint integral operator defined by

$$K^*g(t) = \int_a^b \overline{k(s,t)}g(s)ds , \quad g \in L_2(a,b) , \quad t \in [a,b] .$$

It seems to me possible that even the latter choice might turn out to be of some practical value, at least if k is symmetric, i.e. $k(t,s) = \overline{k(s,t)}$; for the additional difficulty over the iterated Galerkin method is then perhaps not as great as might be thought. If k is symmetric, then $K^* = K$, and the equations (6.8), with $v_{ni} = K^*u_{ni} = Ku_{ni}$, can be written

6.9 $$\sum_{i=1}^{n} [(u_{nj},Ku_{ni}) - (Ku_{nj},Ku_{ni})]a_{ni} = (u_{nj},Kf) , \quad j = 1,\ldots,n .$$

The first term of the left-hand side matrix is already required in the Galerkin method. The second can be evaluated as

6.10 $$(Ku_{nj},Ku_{ni}) = (z_{nj},z_{ni}) \ ,$$

where $$z_{ni}(t) = \int_a^b k(t,s)u_{ni}(s)ds \ ,$$

whereas the first term is just

$$(u_{nj},Ku_{ni}) = (u_{nj},z_{ni}) \ .$$

In most applications of the Galerkin method, the latter inner products will have to be evaluated numerically. That being the case, the Galerkin method itself requires the calculation of numerical values of $z_{ni}(t)$ for t ranging over a suitable set of quadrature points, and there is then almost no extra cost, and little inconvenience, in simultaneously calculating the inner products (6.10). However, the linear equations (6.9) do have one definite disadvantage compared with the Galerkin equations (6.4): the matrix in (6.9) can easily turn out to be rather poorly conditioned; and this could in turn lead to a need for more accurate inner products. It could be interesting to see some experimentation, perhaps starting from Chandler's version of the Galerkin method, with splines and numerical quadratures.

A convergence proof for the second choice of v_{ni}, restricted to the case in which k is symmetric and positive-definite, has been given in [38]. The theory in [38], and also numerical experience in [41], both suggest that for the positive-definite case this method should generally be at least as accurate as the iterated Galerkin method, and sometimes more accurate, particularly if the equation has a characteristic value near 1.

7. Conclusion

I hope I have given you some idea of the rich variety of methods that exist for Fredholm integral equations of the second kind, and some appreciation of the beautiful interplay that exists in this field between profound mathematics on the one hand, and down-to-earth practical matters on the other. I hope, too, that I have left you with the feeling that the subject is by no means a closed book; that in fact there is ample room for improvement in both theory and practice.

Of the methods not covered in this review, the omission I most regret is the Padé approximant technique (see [14,23]). This is a method that seems little known among numerical analysts, and the numerical analysis of the method is not yet complete. Yet it is a method that can be used, as in the work of Kloet and Tjon [28], to tackle problems of the most extreme computational difficulty. All I can do at this stage, however, is to mention the method and direct you to the literature.

References

[1] Anselone, P.M. Collectively Compact Operator Approximation Theory and Applications to Integral Equations, Prentice-Hall, Englewood Cliffs, N.J., 1971.

[2] Atkinson, K.E. Extensions of the Nyström method for the numerical solution of linear integral equations of the second kind, U.S. Army M.R.C. Technical Summary Rep. #686, Madison, Wisconsin, 1966.

[3] Atkinson, K.E. The numerical solution of Fredholm integral equations of the second kind, *SIAM, J. Numer. Anal.*, 4, 337-348, 1967.

[4] Atkinson, K.E. Iterative variants of the Nystrom method for the numerical solution of integral equations, *Numer. Math.*, 22, 17-31, 1973.

[5] Atkinson, K.E. A Survey of Numerical Methods for the Solution of Fredholm Integral Equations of the Second Kind, SIAM, Philadelphia, 1976.

[6] Atkinson, K.E. An automatic program for linear Fredholm integral equations of the second kind, *ACM Trans. on Math. Software*, 2, 154-171 and 196-199, 1976.

[7] Baker, C.T.H. Expansion methods, *Numerical Solution of Integral Equations* (ed. L.M. Delves and J. Walsh), Clarendon Press, Oxford, 80-96, 1974.

[8] Baker, C.T.H. The Numerical Treatment of Integral Equations, Clarendon Press, Oxford, 1977.

[9] Bateman, H. On the numerical solution of linear integral equations, *Proc. Roy. Soc. London Ser. A*, 100, 441-449, 1922.

[10] Brakhage, H. Über die numerische Behandlung von Integralgleichungen nach der Quadraturformelmethode, *Numer. Math*, 2, 183-196, 1960.

[11] Buckner, H. Die Praktische Beihandlung von Integralgleichungen, Springer-Verlag OHG, Berlin, 1952.

[12] Buckner, H. Numerical methods for integral equations, in *Survey of Numerical Analysis* (ed. J. Todd), McGraw-Hill, 439-467, 1962.

[13] Chandler, G.A. Global superconvergence of iterated Galerkin solutions for second kind integral equations, submitted for publication.

[14] Chisholm, J.S.R. Padé approximants and linear integral equations, in *The Padé Approximant in Theoretical Physics* (ed. G.A. Baker Jr. and J.L. Gammel), Academic Press, New York, 171-182, 1970.

[15] Clenshaw, C.W. and Curis, A.R. A method for numerical integration on an automatic computer, *Numer. Math.* 2, 197-205, 1960.

[16] Davis, P.J. Interpolation and Approximation, Blaisdell, Waltham, Massachusetts, 1963.

[17] Delves, L.M. A fast method for the solution of Fredholm integral equations, *J. Inst. Maths. Applics.*, 20, 173-182, 1977.

[18] Delves, L.M. and Walsh J. (eds.) Numerical Solution of Integral Equations, Clarendon Press, Oxford, 1974.

[19] El-gendi, S.E. Chebyshev soltuion of differential, integral and integro-differential equations, *Computer J.*, 12, 282-287, 1969.

[20] Elliott, D. A Chebyshev series method for the numerical solution of Fredholm integral equations, *Computer J.*, 6, 102-111, 1963.

[21] Elliott, D. and Warne, W.G. An algorithm for the numerical solution of linear integral equations, *Internat. Computation Centre Bull.*, 6, 207-224, 1967.

[22] Graham, I.G. and Sloan, I.H. On the compactness of certain integral operators, *J. Math. Anal. Applics.*, 68, 580-594, 1979.

[23] Graves-Morris, P.R. Padé approximants for linear integral equations?, *J. Inst. Maths. Applics.*, 21, 375-378, 1978.

[24] Houstis, E.N. and Papatheodorou, T.S. A collocation method for Fredholm integral equations of the second kind, *Math. Comp.*, 32, 159-173, 1978.

[25] Ikebe, Y. The Galerkin method for the numerical solution of Fredholm integral equations of the second kind, *SIAM Review*, 14, 465-491, 1972.

[26] Imhof, J.P. On the method for numerical integration of Clenshaw and Curtis, *Numer. Math.*, 5, 138-141, 1963.

[27] Kantorovich, L.V. and Akilov, G.P. Functional Analysis in Normed Spaces, Pergamon Press, Oxford, 1964.

[28] Kloet, W.M. and Tjon, J.A. Elastic neutron-deuteron scattering with local potentials, *Annals of Physics*, 79, 407-440, 1973.

[29] Marsh, T. and Wadsworth, M. An iterative method for the solution of Fredholm integral equations of the second kind, *J. Inst. Maths. Applics.* 18, 57-65, 1976.

[30] Mayers, D.F. Quadrature methods for equations of the second kind, in *Numerical Solution of Integral Equations* (ed. L.M. Delves and J. Walsh), Clarendon Press, Oxford, 64-79, 1974

[31] Mikhlin S.G. and Smolitskiy, K.L. Approximate Methods for Solution of Differential and Integral Equations, American Elsevier, New York, 1967.

[32] Noble, B. A bibliography on methods for solving integral equations, *Mathematics Research Center Tech. Reps.* #1176 (author listing) and #1177 (subject listing), Madison, Wisconsin, 1971.

[33] Nyström, E.J. Über die praktische Auflosung von Integralgleichungen mit Anwendungen auf Randwertaufgaben, *Acta Math.*, 54, 185-204, 1930.

[34] Phillips, J.L. The use of collocation as a projection method for solving linear operator equations, *SIAM J. Numer. Anal.*, 9, 14-28, 1972.

[35] Piessens, R. and Branders, M. Numerical solution of integral equations of mathematical physics, using Chebyshev polynomials, *J. Comp. Phys.* 21, 178-196, 1976.

[36] Prenter, P. A collocation method for the numerical solution of integral equations, *SIAM J. Numer. Anal.*, 10, 570-581, 1973.

[37] Sloan, I.H. Error analysis for a class of degenerate-kernel methods, *Numer. Math.*, 25, 231-238, 1976.

[38] Sloan, I.H. Convergence of degenerate-kernel methods, *J. Austral. Math. Soc., Series B*, 19, 422-431, 1976.

[39] Sloan, I.H. On the numerical evaluation of singular integrals, *BIT*, 18, 91-102, 1978.

[40] Sloan, I.H. and Burn, B.J. Collocation with polynomials for integral equations of the second kind: a new approach to the theory, *J. Integral Equations*, 1, 77-94, 1979.

[41] Sloan, I.H., Burn, B.J. and Datyner, N. A new approach to the numerical solution of integral equations, *J. Comp. Phys.*, 18, 92-105, 1975.

[42] Sloan, I.H., Noussair, E. and Burn, B.J. Projection methods for equations of the second kind, *J. Math. Anal. Applics.*, 69, 84-103, 1979.

[43] Sloan, I.H. and Smith W.E. Product integration with the Clenshaw-Curtis points: implementation and error estimates, submitted for publication.

[44] Thompson, G.T. On Bateman's method for solving linear integral equations, *J. Assoc. for Comput. Mach.*, 4, 314-328, 1957.

[45] Walther, A. and Dejon, B. General report on the numerical treatment of integral and integro-differential equations, in *Symposium on the Numerical Treatment of Ordinary Differential Equations, Birkhäuser Verlag*, Basel/Stuttgart, 645-671, 1960.

[46] Young, A. The application of approximate product integration to the numerical solution of integral equations, *Proc. Roy. Soc. London*, Ser. A, 224, 561-573, 1954.

SOME APPLICATION AREAS FOR FREDHOLM INTEGRAL EQUATIONS OF THE SECOND KIND

Ivan G. Graham
School of Mathematics, University of New South Wales, Sydney, N.S.W.

1. Introduction

In the seventy or so years since its inception, the elegant theory of existence and uniqueness of solutions of second kind integral equations as originally put forward by Fredholm [6] has had many applications. In the early years after the birth of the theory its main applications were of a theoretical nature. For example, in applied mathematics, reformulations of classical boundary value problems as integral equations paved the road towards a unified theory of existence and uniqueness of solutions to these problems. In pure mathematics, integral equations stimulated greatly the early development of functional analysis. Evidence of this can be seen in most modern functional analysis texts where the theory of Fredholm can be found as a special case of the more abstract theory of compact operators (see, for example, [21, chapter 4]).

Although it was known for a long time that many physical problems could be formulated as integral equations, it is only relatively recently that solutions have actually been generated from these formulations. We can conjecture several reasons for the lack of practical use, in the past, of integral equations. Most problems have both a differential equation and an integral equation formulation. The advantages of the integral equation include the fact that it usually involves fewer dimensions than the differential equation. It also has built into it any boundary conditions which the problem must satisfy, and a convenient theory of existence and uniqueness of solutions is available, courtesy of Fredholm.

However, the numerical solution of the integral equation has in the past posed great difficulties, since, if any reasonable accuracy is to be attained, we must solve a fairly large system of linear equations. In the pre-computer era that was a daunting prospect.

On the other hand, the solution of the differential equation formulation of the problem also involves a large scale discretisation process. But, in this case the matrices involved are sparse, and various relaxation techniques allow the problem to be attacked on a step-by-step manner.

The availability of modern computers has gone a long way towards neutralising the traditional handicap of the integral equation method, and in the last fifteen years integral equations have become an important tool, not only in theory, but also in the actual solution of problems. However, much work has still to be done, particularly on theory and numerical analysis, before integral equations can achieve full potential as a practical tool.

In this paper we propose to outline some aspects of both theoretical and practical applications. In Section 3 we describe the use of integral equations in the theory of boundary value problems for Laplace's equation in three-dimensional regions. We demonstrate the method in a rigorous manner and aim not only to point out the power of the Fredholm theory, but also to generate interest in the method itself which is applicable to a much wider class of problems than those chosen. We also discuss some present (and possible future) practical applications of the method, the numerical difficulties that may be encountered, and how to overcome them.

In Section 4 we describe the use of an integral equation method to determine the distribution of current in a conductor with rectangular cross-section. The integral equation for the current distribution is derived from first principles and solved numerically using both the collocation method and a simple iterative variant of it.

Since the methods discussed in Sections 3 and 4 have as their basis the classical results of potential theory, and often make use of the Fredholm Alternative Theorem, we describe these in Section 2.

The recent publication of Jawson and Symm [8] contains much interesting information, both theoretical and practical, on the use of integral equations to solve problems in potential theory. In a broader context the reference works by Noble [14] and [15] provide a wealth of information on theory and applications of integral equations in general, while in [12] Lonseth gives an illuminating discussion of the historical development of integral equations and a review of some general areas of application.

2. Mathematical Basis

(i) *Potential Theory*

The origin of potential theory lies in the investigation of the gravitational field produced by a point of mass m with coordinates $\underset{\sim}{x}' = (x',y',z')$ in space. It was found that the force $\underset{\sim}{F}$ exerted on any other body of unit mass with coordinates $\underset{\sim}{x} = (x,y,z)$ is given by

$$\underset{\sim}{F}(\underset{\sim}{x}) = -\frac{m(\underset{\sim}{x} - \underset{\sim}{x}')}{|\underset{\sim}{x} - \underset{\sim}{x}'|^3}, \quad \text{where } \underset{\sim}{x} \neq \underset{\sim}{x}',$$

and where $|\cdot|$ denotes the usual Euclidean norm

$$|\underset{\sim}{x}| = \sqrt{x^2 + y^2 + z^2} .$$

It is obvious that the function ϕ defined in $\mathbb{R}^3 \setminus \{\underset{\sim}{x}'\}$ by

2.1 $$\phi(\underset{\sim}{x}) = \frac{m}{|\underset{\sim}{x} - \underset{\sim}{x}'|} ,$$

is a scalar potential for $\underset{\sim}{F}$ in the sense that

2.2 $$\underset{\sim}{F} = \nabla\cdot\phi$$

where the differentiation in (2.2) is carried out with respect to the variable $\underset{\sim}{x}$. It is easy to show that ϕ given by (2.1) satisfies Laplace's equation

2.3 $$\nabla^2 u = 0$$

in $\mathbb{R}^3 \setminus \{\underset{\sim}{x}'\}$.

Let S denote a "suitably smooth" (see below) closed bounded surface in $\mathbb{R}^3$. We shall denote an element of area at any point $\underset{\sim}{s}$ (or $\underset{\sim}{s}'$) of S as dS (or dS'), and the directional derivative with respect to the normal to S at $\underset{\sim}{s}$ (or $\underset{\sim}{s}'$) as $\frac{\partial}{\partial n'}$ (or $\frac{\partial}{\partial n'}$). Unless otherwise stated, the normal will be directed outward from S. Then, if σ,μ are continuous functions on S the functions v and w defined by

2.4 $$v(\underset{\sim}{x}) = \int_S \frac{1}{|\underset{\sim}{x} - \underset{\sim}{s}'|}\,\sigma(\underset{\sim}{s}')dS' ,$$

and

2.5 $$w(\underset{\sim}{x}) = \int_S \frac{\partial}{\partial n'}\left(\frac{1}{|\underset{\sim}{x} - \underset{\sim}{s}'|}\right)\mu(\underset{\sim}{s}')dS' ,$$

satisfy (2.3) in $\mathbb{R}^3 \setminus S$.

Of course, for the integrals (2.4) and (2.5) to exist for all $\underset{\sim}{x} \in \mathbb{R}^3$ certain assumptions must be made on the smoothness of the surface S. We shall assume throughout that S is a Lyupanov surface (see [22, p.385]). The existence of the integrals (2.4) and (2.5) then follows [22, pp.384-386]. We let V_i denote the region interior, and V_e the region exterior to S. Then, the following definition is commonly made.

DEFINITION 2.1

The function v (or w) as defined on $\mathbb{R}$ by (2.4) (or (2.5)) is called the potential of a single (or double) layer due to surface distribution σ (or moment μ).

We follow common terminology by referring to any solution u of (2.3) as a harmonic function. Also, if u has continuous derivatives of order up to and including 2, we say that u is a regular function. Then, the following Lemma follows easily, and is stated without proof.

LEMMA 2.2 *The function* v *and* w *defined by (2.4) and (2.5) are regular harmonic functions on* $\mathbb{R}^3 \setminus S$.

The following Lemmas ((2.3) and (2.4)) further describe the properties of v and w, and are also stated without proof. A proof can be found in any standard text on potential theory (see, for example, [20, Chapter V]). First, a little explanation about notation is necessary.

Let $\underset{\sim}{s}$ be a point on S, and let u be any function defined on $\mathbb{R}^3$. Then, we use the notation $u_-(\underset{\sim}{s})$ (or $u_+(\underset{\sim}{s})$) to denote the limit $\lim_{\underset{\sim}{x}\to\underset{\sim}{s}} u(\underset{\sim}{x})$, where the limit is taken as $\underset{\sim}{x}$ tends to $\underset{\sim}{s}$ in V_i (or (V_e)). Similarly we use the notation $\frac{\partial u}{\partial n_-}(\underset{\sim}{s})$ $\left(\text{or } \frac{\partial u}{\partial n_+}(\underset{\sim}{s})\right)$ to denote the limit $\lim_{\underset{\sim}{x}\to\underset{\sim}{s}} \frac{\partial u}{\partial n}(\underset{\sim}{x})$, which is the limit of the derivative of u in the direction of the outward normal at $\underset{\sim}{s}$, taken as $\underset{\sim}{x}$ tends to $\underset{\sim}{s}$ in V_i (or (V_e)).

LEMMA 2.3 *The function* v *is continuous on* $\mathbb{R}^3$, *whereas the function* $\frac{\partial v}{\partial n}$ *satisfies*

2.6 $$\frac{\partial v}{\partial n_-}(\underset{\sim}{s}) = 2\pi\, \sigma(\underset{\sim}{s}) + \frac{\partial v}{\partial n}(\underset{\sim}{s}) ,$$

and

2.7 $$\frac{\partial v}{\partial n_+}(\underset{\sim}{s}) = -\, 2\pi\, \sigma(\underset{\sim}{s}) + \frac{\partial v}{\partial n}(\underset{\sim}{s}) ,$$

for all $\underset{\sim}{s} \in S$.

LEMMA 2.4 *The function* w *satisfies*

2.8 $$w_-(\underset{\sim}{s}) = -\, 2\pi\, \mu(\underset{\sim}{s}) + w(\underset{\sim}{s}) ,$$

and

2.9 $$w_+(\underset{\sim}{s}) = 2\pi\, \mu(\underset{\sim}{s}) + w(\underset{\sim}{s}) ,$$

for all $\underset{\sim}{s} \in S$.

In addition to these basic results for single and double layer potentials, we shall require certain well known properties of harmonic functions. The most important of these is the fact that a function which is regular and harmonic in either of the regions V_i or V_e can be represented in that region as a linear combination of single and double layer potentials. As we shall see, in the case of the infinite region V_e, an extra condition on the function must be satisfied before the representation is valid. Lemma (2.5) provides the basic results needed to construct such a representation.

LEMMA 2.5 (Green's Theorem in the interior domain). *Suppose* u *and* $\bar{u}$ *are continuous and have continuous first derivatives in* $S \cup V_i$ *and are regular in* V_i. *Then*

2.10 $$\int_{V_i} u\nabla^2\bar{u}\, dV + \int_{V_i} \nabla u \cdot \nabla\bar{u}\, dV = \int_S u \frac{\partial\bar{u}}{\partial n}\, dS ,$$

and

2.11 $$\int_{V_i} (\bar{u}\nabla^2 u - u\nabla^2\bar{u})dV = \int_S \left(\bar{u}\frac{\partial u}{\partial n} - u\frac{\partial\bar{u}}{\partial n}\right) dS .$$

Proof: The proof follows easily from Gauss's Divergence Theorem.

COROLLARY 2.6 *Suppose* u *and* $\bar{u}$ *satisfy the conditions of Lemma (2.5) and are also harmonic in* V_i. *Then*

2.12 $$\int_{V_i} \nabla u \cdot \nabla u\, dV = \int_S u \frac{\partial u}{\partial n}\, dS$$

and

2.13 $$\int_S \left(\bar{u}\frac{\partial u}{\partial n} - u\frac{\partial\bar{u}}{\partial n}\right) dS = 0 .$$

Proof: The proof follows immediately from Lemma (2.5).

COROLLARY 2.7 *Suppose* u *and* $\bar{u}$ *are continuous and have continuous first derivatives in* $S \cup V_e$, *and are regular harmonic functions in* V_e. *Suppose also that* $u(\underset{\sim}{x})$, $\bar{u}(\underset{\sim}{x})$ *approach zero uniformly as* $\underset{\sim}{x}\to\infty$. *Then*

2.14 $$\int_{V_e} \nabla u \cdot \nabla u\, dV = \int_S u \frac{\partial u}{\partial n}\, dS$$

and

2.15 $$\int_S \left(\bar{u}\frac{\partial u}{\partial n} - u\frac{\partial\bar{u}}{\partial n}\right) dS = 0 ,$$

where in this case the normal derivatives are inward on S *(i.e. outward from* V_e*).*

Proof: Choose a sphere, centred on the origin, and large enough to contain V_i as a proper subset, and denote its surface by R and its radius by r.

Let T denote the region bounded by S and R, and apply (2.12) to the region $T \cup S \cup R$, to get

2.16 $$\int_T \nabla u \cdot \nabla u \, dV = \int_S u \frac{\partial u}{\partial n} dS + \int_R u \frac{\partial u}{\partial n} dR .$$

Note that in (2.16) the directional derivatives are with respect to the outward normal from T. Now, converting to polar coordinates, we have

$$\int_R u \frac{\partial u}{\partial n} dR = \int_0^{\pi} \int_0^{2\pi} u(r,\theta,\phi) \frac{\partial u}{\partial n} (r,\theta,\phi) r^2 \sin\theta d\theta d\phi ,$$

where r denotes the radius of R. Thus, it follows that

$$\left|\int_R u \frac{\partial u}{\partial n} dR\right| \leq \frac{1}{r} \left|\int_0^{\pi} \int_0^{2\pi} ru(r,\theta,\phi) \; r^2 \frac{\partial u}{\partial n} (r,\theta,\phi) \sin\theta d\theta d\phi\right|$$

$$\to 0 \text{ as } r \to \infty .$$

The limit is obtained since the condition that u is regular harmonic in V_e and $u(\underset{\sim}{x}) \to 0$ uniformly as $\underset{\sim}{x} \to \infty$ is sufficient to ensure that the functions $ru(r,\theta,\phi)$ and $r^2 \frac{\partial u}{\partial n} (r,\theta\,\phi)$ remain bounded as $r \to \infty$ (see [22, p.329]). The relation (2.14) then follows by letting $r \to \infty$ in (2.16).

To get (2.15) we apply (2.13) to the region T, and let $r \to \infty$. In this case we use the fact that $\bar{u}$ tends to zero uniformly at infinity.

LEMMA 2.8 *Suppose* u *is continuous with continuous first derivatives in* $S \cup V_i$, *and suppose* u *is a regular harmonic function in* V_i. *Then*

2.17 $$\int_S \left\{ \frac{1}{|\underset{\sim}{x} - \underset{\sim}{s}'|} \frac{\partial u}{\partial n'} (\underset{\sim}{s}') - u(\underset{\sim}{s}') \frac{\partial}{\partial n'} \left(\frac{1}{|\underset{\sim}{x} - \underset{\sim}{s}'|} \right) \right\} dS' = \begin{Bmatrix} 0 & (\underset{\sim}{x} \in V_e) \\ 2\pi u(\underset{\sim}{x}) & (\underset{\sim}{x} \in S) \\ 4\pi u(\underset{\sim}{x}) & (\underset{\sim}{x} \in V_i) \end{Bmatrix} .$$

Proof: For the case when $\underset{\sim}{x} \in V_e$, define $\bar{u}$ on $V_i \cup S$ by

2.18 $$\bar{u}(\underset{\sim}{x}') = \frac{1}{|\underset{\sim}{x} - \underset{\sim}{x}'|}$$

and the result follows by a simple application of Corollary (2.6).

Suppose now that $\underset{\sim}{x} \in V_i$. Choose a sphere centred on $\underset{\sim}{x}$ with radius sufficiently small so that it is a proper subset of V_i, and denote its surface by H and its radius by h. Define $\bar{u}$ in the closed region bounded by $S \cup H$ by (2.18), and apply Corollary (2.6) to u and $\bar{u}$ in that region to obtain

$$\int_S \left\{ \frac{1}{|\underset{\sim}{x} - \underset{\sim}{s}'|} \frac{\partial u}{\partial n'}(\underset{\sim}{s}') - u(\underset{\sim}{s}') \frac{\partial}{\partial n'} \left(\frac{1}{|\underset{\sim}{x} - \underset{\sim}{s}'|} \right) \right\} dS' =$$

2.19

$$-\int_H \left\{ \frac{1}{|\underset{\sim}{x} - \underset{\sim}{x}'|} \frac{\partial u}{\partial n'}(\underset{\sim}{x}') - u(\underset{\sim}{x}') \frac{\partial}{\partial n'} \frac{1}{|\underset{\sim}{x} - \underset{\sim}{x}'|} \right\} dH' .$$

Now, when $\underset{\sim}{x}' \in H$, the outward normal to the region $S \cup H$ at that point is in the direction $(\underset{\sim}{x} - \underset{\sim}{x}')$. Also, $|\underset{\sim}{x} - \underset{\sim}{x}'| = h$, and it follows that

2.20
$$\int_H u(\underset{\sim}{x}') \frac{\partial}{\partial n'} \left(\frac{1}{|\underset{\sim}{x} - \underset{\sim}{x}'|} \right) dH' = \frac{1}{h^2} \int_H u(\underset{\sim}{x}')\ dH' ,$$

and

2.21
$$\int_H \frac{1}{|\underset{\sim}{x} - \underset{\sim}{x}'|} \frac{\partial u}{\partial n'}(\underset{\sim}{x}')\ dH' = \frac{1}{h} \int_H \frac{\partial u}{\partial n'}(\underset{\sim}{x}')\ dH' .$$

Using the mean value theorem for harmonic functions, (see [22, p.320]), (2.20) becomes

2.20'
$$\int_H u(\underset{\sim}{x}') \frac{\partial}{\partial n'} \left(\frac{1}{|\underset{\sim}{x} - \underset{\sim}{x}'|} \right) dH' = 4\pi u(\underset{\sim}{x}) ,$$

and from (2.21), since $\frac{\partial u}{\partial n'}$ is continuous on H, we obtain

2.21'
$$\lim_{h\to 0} \int_H \frac{1}{|\underset{\sim}{x} - \underset{\sim}{x}'|} \frac{\partial u}{\partial n'}(\underset{\sim}{x}')\ dH' = \lim_{h\to 0} \frac{1}{h} \int_H \frac{\partial u}{\partial n'}(\underset{\sim}{x}')\ dH' = 0 .$$

Thus letting $h\to 0$ in (2.19) and using (2.20)' and (2.21)' we obtain the required result for $\underset{\sim}{x} \in V_i$.

The result for $\underset{\sim}{x} \in S$ is proved similarly, except that in this case we apply Corollary (2.6) to the region bounded by $S_1 \cup H_1$, where H_1 is that portion of the sphere H lying inside V_i, and S_1 is that portion of S lying outside H.

LEMMA 2.9 *Suppose that* u *satisfies the conditions of Lemma (2.8) in* V_e *and* $u(\underset{\sim}{x})\to 0$ *uniformly as* $\underset{\sim}{x}\to\infty$. *Then*

2.22
$$\int_S \left\{ \frac{1}{|\underset{\sim}{x} - \underset{\sim}{s}'|} \frac{\partial u}{\partial n'}(\underset{\sim}{s}') - u(\underset{\sim}{s}') \frac{\partial}{\partial n'} \left(\frac{1}{|\underset{\sim}{x} - \underset{\sim}{s}'|} \right) \right\} dS' = \begin{Bmatrix} 0 & (\underset{\sim}{x} \in V_i) \\ 2\pi u(\underset{\sim}{x}) & (\underset{\sim}{x} \in S) \\ 4\pi u(\underset{\sim}{x}) & (\underset{\sim}{x} \in V_e) \end{Bmatrix}$$

where the normal derivatives are inward on S.

Proof: The proof is analogous to that of Lemma (2.8), except that, since we consider infinite regions, we use Corollary (2.7) instead of Corollary (2.6).

Discussion (2.10), Lemma (2.8) and (2.9) demonstrate how a function, harmonic in a region, and satisfying certain regularity conditions, can be represented in that region as a sum of single and double layer potentials defined on its boundary. However, if we arbitrarily specify continuous functions g_1 and g_2 on S, and consider the integral

$$\int_S \left\{\frac{1}{|\underset{\sim}{x} - \underset{\sim}{s}'|} g_1(\underset{\sim}{s}') - g_2(\underset{\sim}{s}') \frac{\partial}{\partial n'} \left(\frac{1}{|\underset{\sim}{x} - \underset{\sim}{s}'|}\right)\right\} dS' ,$$

then this integral does not, in general, tend to g_2 as $\underset{\sim}{x}$ tends to S. We shall see in Section 3 that the specification of either the value of the function or its normal derivative on the boundary will determine the function uniquely.

In the above development we have been concerned with solutions of Laplace's equation in 3-dimensional regions. It should also be pointed out that the corresponding results in 2-dimensional regions follow a similar pattern except that the basic potentials (2.4) and (2.5) have $1/|\underset{\sim}{x} - \underset{\sim}{x}'|$ replaced by $\ln (1/|\underset{\sim}{x} - \underset{\sim}{x}'|)$ (see [20]).

An obvious generalisation of (2.3) is the Helmholtz equation

2.23 $$(\nabla^2 + \alpha^2)u = 0 \quad (\alpha \geq 0) .$$

It is easy to show that a fundamental solution of (2.23) is

$$\phi_\alpha = \frac{e^{i\alpha|\underset{\sim}{x} - \underset{\sim}{x}'|}}{|\underset{\sim}{x} - \underset{\sim}{x}'|} ,$$

and that the single and double layer Helmholtz potentials defined on $\mathbb{R}^3$ by

2.24 $$v_\alpha(\underset{\sim}{x}) = \int_S \frac{e^{i\alpha|\underset{\sim}{x} - \underset{\sim}{s}'|}}{|\underset{\sim}{x} - \underset{\sim}{s}'|} \sigma(\underset{\sim}{s}')\, dS'$$

and

2.25 $$w_\alpha(\underset{\sim}{x}) = \int_S \frac{\partial}{\partial n'} \frac{e^{i\alpha|\underset{\sim}{x} - \underset{\sim}{x}'|}}{|\underset{\sim}{x} - \underset{\sim}{s}'|} \mu(\underset{\sim}{s}')\, dS'$$

both satisfy (2.23) in the region $\mathbb{R}^3 \setminus S$.

Potential theory methods, similar to those described in the following section, and leading to integral equation formulations of physical problems, are well represented in the literature. Jawson and Symm [8], describe the use of such methods in two and three dimensional regions and their applications to problems in many areas including aerodynamics, heat conduction and electrostatics.

For the more specific case of boundary value problems for the Helmholtz equation, Kleinman and Roach [11] and Burton [5], both use the potentials (2.24) and (2.25) to re-formulate those problems as integral equations. Their work has physical applications in acoustic and electromagnetic scattering theory (see [5, pp.6-14]).

(ii) *Fredholm Theory*

The Fredholm Alternative Theorem provides valuable information on the solution y, of the integral equation

2.26
$$y(\underset{\sim}{s}) = p(\underset{\sim}{s}) + \lambda \int_\Omega k(\underset{\sim}{s},\underset{\sim}{s}')y(\underset{\sim}{s}')d\Omega(\underset{\sim}{s}') \ , \quad \underset{\sim}{s} \in \Omega \ ,$$

where Ω is some compact subset of $\mathbf{R}^n$, k (the kernel function) and p (the inhomogeneous term) are given complex valued functions defined on $\Omega \times \Omega$ and Ω respectively, and λ is a given complex number. Here and below $d\Omega(\underset{\sim}{s}')$ denotes an element of Ω at $\underset{\sim}{s}'$.

DEFINITION 2.11 Given equation (2.26) we define the adjoint equation to be

2.27
$$z(\underset{\sim}{s}) = q(\underset{\sim}{s}) + \bar{\lambda} \int_\Omega \overline{k(\underset{\sim}{s}',\underset{\sim}{s})}\, z(\underset{\sim}{s}')d\Omega(\underset{\sim}{s}') \ , \quad \underset{\sim}{s} \in \Omega \ ,$$

where $\bar{\alpha}$ denotes the complex conjugate of any complex number α, z is the solution to be determined and q is a given complex-valued function on Ω.

For brevity, we shall define the integral operators K and K* by

2.28
$$Ky(\underset{\sim}{s}) = \int_\Omega k(\underset{\sim}{s},\underset{\sim}{s}')y(\underset{\sim}{s}')d\Omega(\underset{\sim}{s}') \ , \quad \underset{\sim}{s} \in \Omega \ ,$$

and

2.29
$$K^*z(\underset{\sim}{s}) = \int_\Omega \overline{k(\underset{\sim}{s}',\underset{\sim}{s})}\, z(\underset{\sim}{s}')d\Omega(\underset{\sim}{s}') \ , \quad \underset{\sim}{s} \in \Omega \ ,$$

for each function y and z defined on Ω, and we shall refer to K as the integral operator induced by k. Using operator notation, we rewrite (2.26) and (2.27) in the following way

2.30
$$y = p + \lambda Ky \ ,$$

and

2.31
$$z = q + \bar{\lambda}K^*z \ ,$$

and we also define the corresponding homogeneous equations to be

2.32 $$y = \lambda K y \ ,$$

and

2.33 $$z = \bar{\lambda} K^* z \ .$$

We shall let $L^2(\Omega)$ denote the space of all complex-valued square integrable functions defined on Ω, equipped with the norm

$$\|y\|_2 = \left\{\int_\Omega |y(\underset{\sim}{s})|^2 d\Omega(\underset{\sim}{s})\right\}$$

and let $C(\Omega)$ denote the space of all complex-valued continuous functions defined on Ω equipped with the norm

$$\|y\|_\infty = \sup_{s \in \Omega} |y(\underset{\sim}{s})| \ .$$

Both $L^2(\Omega)$ and $C(\Omega)$ are Banach spaces of functions, i.e. they are complete in the mathematical sense (see [21, p.74, p.98]). Our version of the Fredholm Alternative will give information about the solutions of (2.26) (or equivalently (2.30)), whenever the integral operator K has certain properties. In order to encapsulate these properties we make the following definition.

DEFINITION 2.12 An operator K on a Banach space X is called compact if it is bounded and maps bounded sets into sets with compact closure.

THEOREM 2.13 (The Fredholm Alternative in $L^2(\Omega)$). *Suppose* K *given by (2.28) is a compact operator from* $L^2(\Omega)$ *to* $L^2(\Omega)$. *Then either Case I or Case II given below holds.*

Case I. The equations (2.30) and (2.31) both have unique solutions for every $p,q, \in L^2(\Omega)$.

Case II. The equations (2.32) and (2.33) have the same number (n say, where $n \geq 1$) of linearly independent solutions in $L^2(\Omega)$.

In the event that Case II holds, let $y_1,\ldots,y_n$ be the solutions of (2.32), and let $z_1,\ldots,z_n$ be the solutions of (2.33). Then (2.30) has a (non-unique) solution in $L^2(\Omega)$ if, and only if $\int_\Omega p z_i = 0$ for each $i = 1,\ldots,n$, and (2.31) has a (non-unique) solution in $L^2(\Omega)$ if, and only if $\int_\Omega q\, y_i = 0$ for each $i = 1,\ldots,n$.

Proof: Most books on integral equations contain some version of the Fredholm Alternative. The above theorem is a special case of the more general result for Banach spaces proved by Kantorovich and Akilov [9, p.497].

Theorem (2.13) is a convenient starting point for an existence theory of solutions to (2.30). In order to verify that K is compact from $L^2(\Omega)$ to $L^2(\Omega)$ the kernel must be shown to satisfy certain conditions, a task which is sometimes difficult in non-standard cases. However, the operators which we shall be dealing with in this paper are of a rather special nature. They all have kernel functions defined on $S \times S$ (where S is the surface described in part (i) of this section) and they satisfy

2.34
$$k(\underset{\sim}{s},\underset{\sim}{s}') = \frac{B(\underset{\sim}{s},\underset{\sim}{s}')}{|\underset{\sim}{s} - \underset{\sim}{s}'|^{2-\delta}} \;, \quad \underset{\sim}{s},\underset{\sim}{s}' \in S \;,$$

where $B(\underset{\sim}{s},\underset{\sim}{s}')$ is bounded on $S \times S$ and continuous for $\underset{\sim}{s} \neq \underset{\sim}{s}'$ and $1 > \delta > 0$. These are in the class of well studied kernels of potential type and it can be shown (see [9, p.537]) that the integral operator induced by k is compact from $L^2(S)$ to $L^2(S)$. It can also be shown that K maps C(S) to C(S), that a certain power of K maps $L^2(S)$ into C(S), and that K* has the same properties as K. These facts can be proved using the methods of [9] and [7].

Since the integral equations of this paper will be studied in the space C(S), we give the following more appropriate version of Theorem (2.13).

COROLLARY 2.14 *Suppose* k *satisfies condition (2.34), and define* K,K* *by replacing* Ω *by* S *in (2.28) and (2.29). Then, considering the equations (2.30)-(2.33) defined for functions acting on* S, *either Case I or Case II given below holds.*

Case I. The equations (2.30) and (2.31) have unique solutions in C(S) for every p and $q \in C(S)$.

Case II. The equations (2.32) and (2.33) have the same number (n say, where $n \geq 1$) of linearly independent solutions in C(S).

In the event the Case II holds, let $y_1,\dots,y_n$ be the solutions of (2.32), and let $z_1,\dots,z_n$ be the solutions of (2.33). Then, (2.30) has a (non-unique) solution in C(S) if, and only if $\int_\Omega p\, z_i = 0$, for each $i = 1,\dots,n$ and (2.31) has a (non-unique) solution in C(S) if, and only if $\int q\, y_i = 0$ for each $i = 1,\dots,n$.

Proof: The proof can be deduced from the properties of the operators K,K* described above, and Theorem (2.13).

3. Boundary Value Problems

In this section we consider a class of boundary value problems for Laplace's equation

2.3
$$\nabla^2 u = 0$$

in three-dimensional regions. We first indicate the uniqueness results for these problems. Then, using the potential theory results of Section 2, we reformulate each problem as an integral equation of the second kind, and use the Fredholm Alternative to show that, provided certain conditions hold, a solution exists, and can be found by solving an integral equation. In our use of potential theory to reformulate problems as integral equations we adopt only the "layer approach" - where the solution is assumed to be either a single or double layer potential. An alternative approach which is well known uses the Green's theorem representations given in Lemmas (2.8) and (2.9). The recent paper by Ahner [1] describes some interesting connections between the two methods.

The integral equation method described here, although classical, is of current research interest. On the theoretical side, for example Roach [16] has used a similar approach to obtain approximate solutions to a general class of boundary value problems for elliptic self adjoint differential equations in n-dimensional spaces.

The method has its most practical relevance, however, when it is used to attack boundary value problems for either Laplace's equation or the Helmholtz equation.

Before defining the boundary value problems we motivate the theory by two simple physical examples.

Example 1. Suppose a body, cubic in shape, and composed of a conducting medium, carries an electric charge. Suppose a steady state has been reached and we wish to determine the electric field $\underset{\sim}{E}$ due to the charge on the body. Then, since $\underset{\sim}{E}$ is irrotational, there exists a scalar function u such that

$$\text{(i)} \qquad \nabla u = \underset{\sim}{E} \, .$$

It follows (see [17, p.32]), that

$$\nabla^2 u = 0$$

in the region exterior to the body. Also, since there is no movement of charge, we have

$$u = \text{constant}$$

on the surface of the body. If we solve this boundary value problem for u, then $\underset{\sim}{E}$ can be determined using (i).

Example 2. (see [5, pp.6-9]). Consider the propagation of small-amplitude acoustic waves in a gas of negligible viscosity. Let ρ denote the density of the gas at rest and let p be the excess pressure and $\underset{\sim}{v}$ be the velocity of the gas at time t. Then $\underset{\sim}{v}$ and p satisfy Newton's equation of motion

$$\text{(ii)} \qquad \rho \frac{\partial \underset{\sim}{v}}{\partial t} = -\nabla p \, .$$

Since there are no sources or sinks, the principle of continuity of mass holds, which, on adoption of standard assumptions about the nature of the gas, reduces to

$$\text{(iii)} \qquad \frac{\partial p}{\partial t} = -\rho\, c^2\, \nabla \cdot \underset{\sim}{v} ,$$

where c is a constant. The motion is irrotational, and so we seek a scalar potential ψ for the velocity $\underset{\sim}{v}$, such that

$$\text{(iv)} \qquad \underset{\sim}{v} = \nabla\psi .$$

By substitution of (iv) into (ii) and integrating, the correct choice of the constant of integration yields

$$\text{(v)} \qquad -\rho\, \frac{\partial\psi}{\partial t} = p ,$$

and substitution of (iv) and (v) into (iii) yields the wave equation

$$\text{(vi)} \qquad \nabla^2\psi - \frac{1}{c^2}\frac{\partial^2\psi}{\partial t^2} = 0 .$$

Now, if we assume that ψ has a harmonic time dependence

$$\text{(vii)} \qquad \psi = ue^{-i\omega t} ,$$

then differentiation with respect to t and substitution into (vi) yields

$$\text{(viii)} \qquad \left(\nabla^2 + \frac{\omega^2}{c^2}\right)u = 0 .$$

This is the Helmholtz equation (2.23) with $\alpha = \frac{\omega}{c}$.

Now, suppose that incident waves are being scattered by a body. Then, when a steady state has been reached, the time independent component u of the velocity potential of the total wave (i.e. the incident wave plus the scattered wave) satisfies (viii) in the region exterior to the scatterer. The problem of finding u is thus a boundary value problem for the Helmholtz equation in the exterior region. The boundary conditions to be imposed depend on the assumptions made about the surface of the scattering body. For a "perfectly soft" scatterer, the excess pressure p is zero at the surface and hence, by (v) and (vii)

$$u = 0, \text{ on the surface of the body.}$$

For a "perfectly hard" scatterer, the normal component of the velocity of the total wave is zero, and it follows from (iv) and (vii) that

$$\frac{\partial u}{\partial n} = 0, \text{ on the surface of the body.}$$

Most practical materials have properties somewhere between that of a hard and soft scatterer, and the boundary condition is of the form

$$\frac{\partial u}{\partial n} + hu = 0, \text{ on the surface of the body.}$$

For the solution of problems in exterior regions, a certain restriction must be made on the behaviour of the unknown function as the field point gets near infinity. This is known as the radiation condition (see [23]).

Now consider the surface S and the regions V_i and V_e as described in Section 2. Let f and h be continuous functions on S, and suppose h is strictly positive. The boundary value problems which we shall consider are given in Table 1.

TABLE 1

Find u such that	In the region	Subject to the conditions	Name of problem
$\nabla^2 u = 0$, u regular	V_i	u continuous on $V_i \cup S$ $u = f$ on S	Interior Dirichlet (3.1)
"	V_i	$u, \frac{\partial u}{\partial n}$ continuous on $V_i \cup S$ $\frac{\partial u}{\partial n} = f$ on S	Interior Neumann (3.2)
"	V_i	$u, \frac{\partial u}{\partial n}$ continuous on $V_i \cup S$ $\frac{\partial u}{\partial n} + hu = f$ on S	Interior Robin (3.3)
"	V_e	u continuous on $V_e \cup S$ $u \to 0$ uniformly at infinity $u = f$ on S	Exterior Dirichlet (3.6)
"	V_e	$u, \frac{\partial u}{\partial n}$ continuous on $V_e \cup S$ $u \to 0$ uniformly at infinity $\frac{\partial u}{\partial n} = f$ on S	Exterior Neumann (3.4)
"	V_e	$u, \frac{\partial u}{\partial n}$ continuous on $V_e \cup S$ $u \to 0$ uniformly at infinity $\frac{\partial u}{\partial n} + hu = f$ on S	Exterior Robin (3.5)

Uniqueness results for the solutions of problems (3.1)-(3.6) follow from the elementary properties of harmonic functions. It can be shown (see [20,pp.183-187]) that each of these problems with the exception of (3.2) has at most one solution, and that (3.2) has at most one solution except for an aribtrary constant.

We have shown in Section 2 that the single and double layer potentials defined by (2.4) and (2.5) are regular harmonic functions in both V_i and V_e. Conversely, Lemmas (2.8) and (2.9) indicate that a function u which is regular and harmonic in either of the regions V_i or V_e can be represented there as a linear combination of a single layer potential with distribution $\frac{\partial u}{\partial n}$ and a double layer potential with moment u. (An extra condition that $u \to 0$ uniformly at infinity is necessary for the representation in V_e). However, the uniqueness results for problems (3.1)-(3.6) imply that such a function u is uniquely determined in the region and its boundary (except, possibly, for an arbitrary constant) by specifying only its value or the value of its normal derivative on the boundary. Thus our strategy for reformulating problems (3.1)-(3.6) will have as its central theme the assumption that the solution can be written as either a single or a double layer potential.

The fact that the Fredholm theory applies to second kind equations influences our decision whether to use a single or double layer potential representation for the solution.

Suppose we consider the interior Dirichlet problem, and assume the solution u can be written as the single layer potential with distribution σ given by (2.4). That is,

3.7 $$u(\underset{\sim}{x}) = v(\underset{\sim}{x}) = \int_S \frac{1}{|\underset{\sim}{x} - \underset{\sim}{s}'|} \sigma(\underset{\sim}{s}')dS' , \quad \underset{\sim}{x} \in V_i .$$

Now, v is continuous on $\mathbb{R}^3$ (Lemma (2.3)) and so, in order to satisfy the boundary conditions of (3.1), the relation

$$f(\underset{\sim}{s}) = \int_S \frac{1}{|\underset{\sim}{s} - \underset{\sim}{s}'|} \sigma(\underset{\sim}{s}')dS' , \quad \underset{\sim}{s} \in S ,$$

must hold. This equation is not of a suitable form for the Fredholm theory to apply, since the unknown σ occurs only once and is under the integral sign. (This is called an integral equation of the first kind). However, if instead we assume the solution u can be written as the double layer potential given by (2.5)

3.8 $$u(\underset{\sim}{x}) = w(\underset{\sim}{x}) = \int_S \frac{\partial}{\partial n'} \left(\frac{1}{|\underset{\sim}{x} - \underset{\sim}{s}'|}\right) \mu(\underset{\sim}{s}')dS' , \quad \underset{\sim}{x} \in V_i ,$$

then, in order to satisfy the boundary conditions of (3.1) we use Lemma (2.4) to obtain the relation

3.9 $$\mu(\underset{\sim}{s}) = -\frac{1}{2\pi} f(\underset{\sim}{s}) + \frac{1}{2\pi} \int_S \frac{\partial}{\partial n'} \left(\frac{1}{|\underset{\sim}{s} - \underset{\sim}{s}'|}\right) \mu(\underset{\sim}{s}')dS' , \quad \underset{\sim}{s} \in S ,$$

an integral equation of the second kind to be solved for μ.

Similarly, our preference for second kind integral equations leads us to choose the single layer representation given by (3.7) for the solution to the interior Neumann problem. To satisfy the boundary conditions for this problem, we use Lemma (2.3) and obtain the second equation

3.10 $$\sigma(\underset{\sim}{s}) = \frac{1}{2\pi} f(\underset{\sim}{s}) - \frac{1}{2\pi} \int_S \frac{\partial}{\partial n} \left(\frac{1}{|\underset{\sim}{s} - \underset{\sim}{s}'|} \right) \sigma(\underset{\sim}{s}')dS' , \quad \underset{\sim}{s} \in S ,$$

to be solved for σ. For the interior Robin problem we again choose a single layer representation for the solution although this is somewhat a matter of taste since a double layer representation also leads to an integral equation of the second kind.

The exterior Neumann and Robin problems are treated similarly. However, if we attempt to apply the above method to the exterior Dirichlet problem (3.6), then our preference for integral equations of the second kind leads us to assume a double layer representation for the solution. However, it is true that (see [20, p.296]) in general, the double layer potential converges to zero at infinity more rapidly than the solution of (3.6). Therefore, the solution of (3.6) cannot always be expressed as a double layer potential, and so other methods of approaching the problem must be used (see [10, pp.313-314]).

The reformulations of problems (3.1)-(3.5) are given in Table 2.

TABLE 2

	Problem	Solution represented as a	Where	Satisfies
3.9	Interior Dirichlet	Double Layer (3.8)	μ	$\mu(\underset{\sim}{s}) = -\frac{1}{2\pi} f(\underset{\sim}{s}) + \frac{1}{2\pi} \int_S \frac{\partial}{\partial n'} \left(\frac{1}{\lvert\underset{\sim}{s} - \underset{\sim}{s}'\rvert} \right) \mu(\underset{\sim}{s}')dS'$
3.10	Interior Neumann	Single Layer (3.7)	σ	$\sigma(\underset{\sim}{s}) = \frac{1}{2\pi} f(\underset{\sim}{s}) - \frac{1}{2\pi} \int_S \frac{\partial}{\partial n} \left(\frac{1}{\lvert\underset{\sim}{s} - \underset{\sim}{s}'\rvert} \right) \sigma(\underset{\sim}{s}')dS'$
3.11	Interior Robin	Single Layer (3.7)	σ	$\sigma(\underset{\sim}{s}) = \frac{1}{2\pi} f(\underset{\sim}{s}) - \frac{1}{2\pi} \int_S \left\{ \frac{\partial}{\partial n} \left(\frac{1}{\lvert\underset{\sim}{s} - \underset{\sim}{s}'\rvert} \right) + \frac{h(\underset{\sim}{s})}{\lvert\underset{\sim}{s} - \underset{\sim}{s}'\rvert} \right\} \sigma(\underset{\sim}{s}')dS'$
3.12	Exterior Neumann	Single Layer (3.7)	σ	$\sigma(\underset{\sim}{s}) = -\frac{1}{2\pi} f(\underset{\sim}{s}) + \frac{1}{2\pi} \int_S \frac{\partial}{\partial n} \left(\frac{1}{\lvert\underset{\sim}{s} - \underset{\sim}{s}'\rvert} \right) \sigma(\underset{\sim}{s}')dS'$
3.13	Exterior Robin	Single Layer (3.7)	σ	$\sigma(\underset{\sim}{s}) = -\frac{1}{2\pi} f(\underset{\sim}{s}) + \frac{1}{2\pi} \int_S \left\{ \frac{\partial}{\partial n} \left(\frac{1}{\lvert\underset{\sim}{s} - \underset{\sim}{s}'\rvert} \right) + \frac{h(\underset{\sim}{s})}{\lvert\underset{\sim}{s} - \underset{\sim}{s}'\rvert} \right\} \sigma(\underset{\sim}{s}')dS'$

The kernel functions of the integral operators in (3.9)-(3.13) all satisfy the condition (2.34). For the functions

$$\frac{\partial}{\partial n'}\left(\frac{1}{|\underset{\sim}{s} - \underset{\sim}{s}'|}\right)$$

and

$$\frac{\partial}{\partial n}\left(\frac{1}{|\underset{\sim}{s} - \underset{\sim}{s}'|}\right) ,$$

this fact follows from the assumed properties of S (see [22, p.387]).

Consequently, since we can write

$$\frac{h(\underset{\sim}{s})}{|\underset{\sim}{s} - \underset{\sim}{s}'|} = \frac{h(\underset{\sim}{s})\ |\underset{\sim}{s} - \underset{\sim}{s}'|^{1-\delta}}{|\underset{\sim}{s} - \underset{\sim}{s}'|^{2-\delta}} , \quad 1 > \delta > 0 ,$$

the kernels of (3.11) and (3.13) also satisfy (2.34).

Thus, we may use Corollary (2.14) to discuss existence and uniqueness of solutions to the integral equations given in Table 2.

THEOREM 3.1 *The interior Dirichlet and the exterior Neumann problems have a solution.*

Proof: The integral equation reformulations of these two problems are given by (3.9) and (3.12) and are an adjoint pair (see Definition (2.11)). Consider the homogeneous version of (3.9) and suppose that there is a non trivial function μ defined on S such that

3.14
$$\mu(\underset{\sim}{s}) = \frac{1}{2\pi}\int_S \frac{\partial}{\partial n'}\left(\frac{1}{|\underset{\sim}{s} - \underset{\sim}{s}'|}\right) \mu(\underset{\sim}{s}')dS' .$$

Then, define w to be the double layer potential with moment μ (see (2.5)). It follows by Lemma (2.4) and equation (3.14) that

$$w_-(\underset{\sim}{s}) = -\ 2\pi\ \mu(\underset{\sim}{s}) + w(\underset{\sim}{s}) = 0 , \quad \underset{\sim}{s} \in S .$$

Thus, w is the solution to the interior Dirichlet problem with zero boundary condition. By the uniqueness of the solution to this problem, we deduce that $w = 0$ in $V_i \cup S$. Thus

$$\frac{\partial w}{\partial n_-}(\underset{\sim}{s}) = 0 , \ \underset{\sim}{s} \in S ,$$

and, since the normal derivative of a double layer satisfies (see [20, p.145])

3.15
$$\frac{\partial w}{\partial n_-}(\underset{\sim}{s}) - \frac{\partial w}{\partial n_+}(\underset{\sim}{s}) = 0 , \quad \underset{\sim}{s} \in S ,$$

it follows that

$$\frac{\partial w}{\partial n_+}(\underset{\sim}{s}) = 0 \ , \quad \underset{\sim}{s} \in S \ ,$$

and hence w is the solution to the exterior Neumann problem with zero boundary condition. By the uniqueness result for this problem, it follows that $w = 0$ in $V_e \cup S$, and hence that

$$w_+(\underset{\sim}{s}) = 0 \ , \quad \underset{\sim}{s} \in S \ .$$

Now, using the relation

$$\mu(\underset{\sim}{s}) = \frac{1}{4\pi}(w_+(\underset{\sim}{s}) - w_-(\underset{\sim}{s})) \ , \quad \underset{\sim}{s} \in S \ ,$$

which follows directly from Lemma (2.4), we obtain

$$\mu(\underset{\sim}{s}) = 0 \ , \quad \underset{\sim}{s} \in S \ .$$

Thus, the equation (3.14) only has the trivial solution, Case I of Corollary (2.14) holds and both equations (3.9) and (3.12) have unique solutions. It follows that the interior Dirichlet and exterior Neumann problems have solutions.

THEOREM 3.2 *The interior Neumann problem has a solution if, and only if the condition*

$$\int_S f(\underset{\sim}{s})dS = 0$$

is satisfied.

Proof: The interior Neumann problem has integral equation reformulation given by (3.10). The adjoint homogeneous equation can be written

$$\mu(\underset{\sim}{s}) = -\frac{1}{2\pi}\int_S \frac{\partial}{\partial n'}\left(\frac{1}{|\underset{\sim}{s} - \underset{\sim}{s}'|}\right)\mu(\underset{\sim}{s}')dS' \ . \tag{3.16}$$

A simple geometric argument (see [20, pp.136-137]) shows that

$$\int_S \frac{\partial}{\partial n'}\left(\frac{1}{|\underset{\sim}{s} - \underset{\sim}{s}'|}\right)dS' = -2\pi \ , \quad \text{for all } \underset{\sim}{s} \in S \ ,$$

which implies that $\mu = 1$ is a non-trivial solution of (3.16). Suppose μ is any other solution of (3.16), and define w to be the double layer with moment μ (see (2.5)). Then, by Lemma (2.4) and equation (3.16), we have

$$w_+(\underset{\sim}{s}) = 2\pi\,\mu(\underset{\sim}{s}) + w(\underset{\sim}{s}) = 0 \ , \quad \underset{\sim}{s} \in S \ .$$

Therefore, w is the solution of the exterior Dirichlet problem with zero boundary value, and thus $w = 0$ in $V_e \cup S$. It follows that

$$\frac{\partial w}{\partial n_+}(\underset{\sim}{s}) = 0 \ , \quad \underset{\sim}{s} \in S \ ,$$

and hence, by (3.15), that

$$\frac{\partial w}{\partial n_-}(\underset{\sim}{s}) = 0 \ , \quad \underset{\sim}{s} \in S \ ,$$

which implies that w is the solution of the interior Neumann problem with zero boundary condition. By the uniqueness result for this problem, it follows that w is a constant in V_i, and so w_- is constant on S. Thus, we can use the relation

$$\mu(\underset{\sim}{s}) = \frac{1}{4\pi}\,(w_+(\underset{\sim}{s}) - w_-(\underset{\sim}{s})) \ , \quad \underset{\sim}{s} \in S \ ,$$

to assert that μ is constant on S. It follows that the only non-trivial solution of (3.16) is a constant, that Case II of Corollary (2.14) holds for (3.10) and, hence, that equation (3.10) has a solution if and only if $\frac{1}{2\pi}\int_S f(\underset{\sim}{s})dS = 0$, and the result follows.

THEOREM 3.3 *The interior and exterior Robin problems have a solution.*

Sketch of Proof: This theorem is proved similarly to Theorem (3.1). The homogeneous versions of (3.11) and (3.13) are shown to have no non-trivial solutions, and the result follows from Corollary (2.14).

DISCUSSION 3.4 The important question of whether the theoretical method described above can be used to actually compute numerical solutions to problems requires some careful consideration. The most important boundary value problems in practice are those given in Table 1, and their obvious generalisations for the Helmholtz equation. Also important are the two dimensional analogues of these problems. It is well known that potential theory methods may be used to reformulate all of these problems as integral equations. However, the advantages of the integral equation formulation, particularly for exterior problems, where a problem in an infinite region is conveniently reduced to a problem on the boundary, may be somewhat offset by computational difficulties.

In general, to solve boundary value problems numerically using this formulation we must first solve an integral equation with a weakly singular kernel over the boundary S (which may be either a one-dimensional or a two-dimensional space). Then, we must calculate, usually by numerical quadrature, the appropriate potential (either single or double layer) using this approximate solution (as the distribution or moment). This two-stage process leaves much room for errors, and to date no rigorous error analysis has been given. However, a method based on a simple piecewise constant collocation method of solution for the integral equation, and a quadrature for finding the potential has been used effectively in practice (see [8, chapter 10]).

In Section 4 a practical illustration of the use of the collocation method to solve a two-dimensional problem is given.

Another important numerical problem arising in potential theory methods, is the case when the integral equation formulation which arises is singular (i.e. the associated homogeneous equation has a non-trivial solution). This is exemplified by the reformulation of the Neumann problem (3.10), which (see Theorem (3.2)) only has a solution if the inhomogeneous term satisfies a certain condition. This problem also arises when potential theory methods are used to reformulate exterior boundary value problems for the Helmholtz equation (2.23). In this case the integral equation for the exterior Dirichlet (Neumann) problem is singular and does not have a unique solution when α is an eigenvalue of the interior Neumann (Dirichlet) problem. For such values of α, although the exterior problem does have a unique solution, the integral equation formulation is not sufficient to generate it. Theoretically, this problem can be solved in a variety of different ways. In [11] Kleinman and Roach use the Helmholtz representation (the Helmholtz analogue of Lemma (2.9)) to obtain a system of integral equations which do have a unique solution. The problem of the numerical solution of integral equations with non-unique solutions has been considered by Lynn and Timlake [13], where the theory of matrix deflations has been used.

This work has particular applications to the forward problem of Electrocardiography (see [3]). This problem involves the construction of a mathematical model of the electrical impulses present within the human body. In particular, it is required to predict, given the impulses produced by the cardiac generator, the electric potential produced at the surface of the body. It can be shown that, given certain assumptions on the conducting properties of the body, the potential satisfies an integral equation with a non-unique solution. This equation is solved numerically in [4].

4. A Numerical Example

When an alternating current flows in a conductor, the magnetic field produced sets up eddy currents which cause the current to be displaced towards the surface of the conductor. This is called the "skin effect" and has been considered by many authors. For example, Silvester [17, pp.227-238] shows that the current distribution in the conductor is the solution to a certain differential equation. However, this problem is not a boundary value problem since we do not stipulate the magnitude of the current at the surface, but rather the total current flowing in the conductor. Since the total current flowing is the integral of the current density vector over the cross-sectional area of the conductor, this problem belongs to the class of problems known as integral value problems.

If the cross section of the conductor is circular, then the differential equation formulation may be solved analytically (see [17, pp.227-228]). However, this is not possible for more general cross-sections and in these cases the problem must be formulated as an integral equation and solved numerically.

In this section we consider a conductor with rectangular cross section, and describe an integral equation method for finding the current distribution. Most of the physical theory presented below can be found in reference [17].

Let $\underset{\sim}{J}$ be the current density vector, and let $\underset{\sim}{E}$ be the electric field intensity. Then, if g is the conductivity of the material we have

$$\underset{\sim}{J} = g\underset{\sim}{E} \ . \tag{4.1}$$

Now, one of Maxwell's equations for time varying fields is

$$\nabla \times \underset{\sim}{E} = -\frac{\partial \underset{\sim}{B}}{\partial t} , \tag{4.2}$$

where $\underset{\sim}{B}$ is the magnetic flux density. Since $\underset{\sim}{B}$ is solenoidal, there exists a vector field $\underset{\sim}{A}$, called the vector potential, satisfying

$$\underset{\sim}{B} = \nabla \times \underset{\sim}{A} \ . \tag{4.3}$$

Substituting (4.3) into (4.2) and integrating yields

$$\underset{\sim}{E} = \underset{\sim}{\nu} - \frac{\partial \underset{\sim}{A}}{\partial t} , \tag{4.4}$$

where $\underset{\sim}{\nu}$ is any irrotational vector, and it follows from (4.1) that the current density vector induced by this electric field is

$$\underset{\sim}{J} = g\underset{\sim}{\nu} - g\frac{\partial \underset{\sim}{A}}{\partial t} \ . \tag{4.5}$$

To use (4.5) to determine $\underset{\sim}{J}$, we must determine the vector potential $\underset{\sim}{A}$ for the magnetic flux of a rectangular conductor.

First consider the field due to a single wire, carrying a steady current j, and choose a cylindrical polar coordinate system so that the current flows in the positive z direction. Then, by symmetry, the magnetic flux density vector at any point outside the wire depends only on r, the perpendicular distance of the point from the wire. Its magnitude is given by the formula (see [17, p.145])

$$\underset{\sim}{B} = \frac{\mu j}{2\pi r} \underset{\sim}{e}_{\theta} ,$$

where μ is the permeability of free space, and a suitable vector potential for this field is given by

$$\underset{\sim}{A} = -\frac{\mu j}{2\pi} [\mathit{ln}\ r + \text{constant}]\hat{\underset{\sim}{z}} , \tag{4.6}$$

where the constant can be chosen arbitrarily to determine the origin of $\underset{\sim}{A}$.

Now, return to the rectangular conductor and suppose that a sinusoidally varying current flows at a low enough angular frequency so that (4.6) remains a good approximaton to the vector potential. Choose a rectangular coordinate system so that the current again flows only in the $\underset{\sim}{z}$ direction. (Since it varies sinusoidally, the current density vector $\underset{\sim}{J}$ may, at any instant be directed in either the positive or negative $\underset{\sim}{z}$ direction). We wish to find the current distribution across the cross section of the conductor at $z = 0$.

Consider an infinitely long thin filament of the conductor with rectangular cross-section $\delta x' \times \delta y'$, running parallel to the z-axis, and passing through the point $(x',y',0)$, (see Figure 1). Then, since in the $z = 0$ plane, assuming harmonic time dependence, we can write

$$\underset{\sim}{J} = J(x',y')\, e^{i\omega t}\, \hat{\underset{\sim}{z}}\,, \tag{4.7}$$

for the current density at the point $(x',y',0)$, it follows from (4.6) that the vector potential due to the filament is given at any other point $(x,y,0)$ by

$$\underset{\sim}{A} = -\frac{\mu}{2\pi}\, e^{i\omega t}\, ln\left(\frac{r}{r_0}\right) J(x',y')\,\delta x'\delta y'\, \hat{\underset{\sim}{z}}\,, \tag{4.8}$$

where $r = \sqrt{(x - x')^2 + (y - y')^2}$, and r_0 is an arbitrary constant which determines the origin of the potential. Thus, the vector potential due to the whole conductor must be given at the point $(x,y,0)$ by

$$\underset{\sim}{A} = -\frac{\mu}{2\pi}\, e^{i\omega t} \int_D ln\left(\frac{\sqrt{(x - x')^2 + (y - y')^2}}{r_0}\right) J(x',y')\,dx'dy'\,\hat{\underset{\sim}{z}}\,, \tag{4.9}$$

where D is the cross-sectional area.

Now, returning to (4.5), we notice that, since $\underset{\sim}{J}$ and $\underset{\sim}{A}$ only have components in the $\underset{\sim}{z}$ direction, so must $\underset{\sim}{\nu}$. It follows from the fact that $\underset{\sim}{\nu}$ is irrotational, that $\underset{\sim}{\nu}$ is independent of x and y and hence in the $z = 0$ plane

$$\underset{\sim}{\nu} = \nu_0\, e^{i\omega t}\, \hat{\underset{\sim}{z}} \tag{4.10}$$

where ν_0 is some constant.

It follows on substitution of (4.7), (4.9) and (4.10) into (4.5), and omitting the vector direction and time-dependent factor that

$$J(x,y) = g\,\nu_0 + \frac{\mu g i\omega}{2\pi} \int_D ln\left(\frac{\sqrt{(x - x')^2 + (y - y')^2}}{r_0}\right) J(x',y')\,dx'dy'\,. \tag{4.11}$$

This is a two-dimensional Fredholm integral equation of the second kind. Recall that the constant r_0 determining the origin of the vector potential can be chosen quite arbitrarily. If we set the total current flowing to a fixed value, then ν_0 in (4.11) will be dependent on

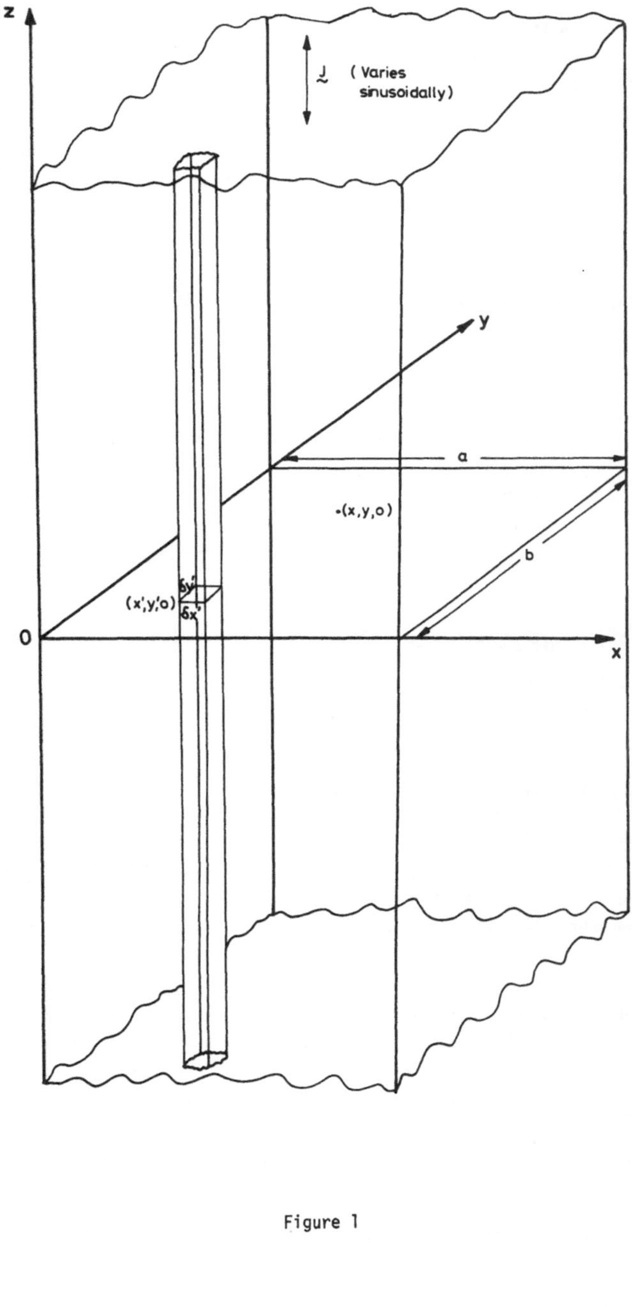

Figure 1

the arbitrarily chosen r_0, and hence will have no physical significance.

Our problem, given that the total current flowing is fixed, is to determine the relative distribution of current throughout the cross-sectional area. If the cross section has length a and breadth b then equation (4.11) (writing the inhomogeneous term as C_0) becomes

4.12 $$J(x,y) = C_0 + \frac{\mu g i \omega}{2\pi} \int_0^b \int_0^a \ln\left(\frac{\sqrt{(x-x')^2 + (y-y')^2}}{r_0}\right) J(x',y')dx'dy' ,$$

and a simple change of scale yields

$$\hat{J}(x,y) = C_0 + \frac{a^2 \mu g i \omega}{2\pi} \int_0^{b/a} \int_0^1 \ln\left(\frac{a\sqrt{(x-x')^2 + (y-y')^2}}{r_0}\right) \hat{J}(x',y')dx'dy' ,$$

where the solution of (4.12) is retreived by the relation

$$J(ax,ay) = \hat{J}(x,y) , \quad (x,y) \in [0,1] \times [0,b/a] .$$

Now, choosing the arbitrary parameter r_0 to take the value a gives the equation

4.13 $$\hat{J}(x,y) = C_0 + \frac{a^2 \mu g i \omega}{2\pi} \int_0^{b/a} \int_0^1 \ln\sqrt{(x-x')^2 + (y-y')^2}\, \hat{J}(x',y')dx'dy' ,$$

Setting $\lambda = \frac{a^2 \mu g i \omega}{2\pi}$, $k(\underset{\sim}{x},\underset{\sim}{x}') = \ln\sqrt{(x-x')^2 + (y-y')^2}$ and $D_0 = [0,1] \times [0,b/a]$, we can write (4.13) more simply as

4.14 $$\hat{J}(\underset{\sim}{x}) = C_0 + \lambda \int_{D_0} k(\underset{\sim}{x},\underset{\sim}{x}')\hat{J}(\underset{\sim}{x}')d\underset{\sim}{x}' , \quad \underset{\sim}{x} \in D_0 ,$$

which is the integral equation we shall solve numerically.

It is a Fredholm integral equation of the second kind. The integral operator is compact when considered as an operator on $C(D_0)$, (see [7]), and, provided the homogeneous equation has no non trivial solution, there exists a unique solution $\hat{J}$ which is continuous on D_0. We shall compute two approximations $\hat{J}_N^I$ and $\hat{J}_N^{II}$ to $\hat{J}$ by using a two-dimensional collocation method.

First, divide D_0 into N sub-rectangles, $\{\Delta_i, i = 1,\ldots,N\}$ and choose the collocation points $\{\underset{\sim}{x}_i : i = 1,\ldots,N\}$ to be the mid-points (i.e. centroids) of each of the rectangles. Then, for each $i = 1,\ldots,N$ define ϕ_i to be the function on D_0 which takes the value 1 on Δ_i and 0 elsewhere. Then, we define $\hat{J}_N^I$ by the expression

4.15 $$\hat{J}_N^I = \sum_{i=1}^N a_i \phi_i ,$$

where the coefficients a_i are chosen so that $\hat{J}_N^I$ satisfies equation (4.14) at each of the collocation points. This leads to the $N \times N$ linear system

4.16 $$a_j = C_0 + \lambda \sum_{i=1}^{N} a_i \int_{\Delta_i} k(\underset{\sim}{x}_j,\underset{\sim}{x}')d\underset{\sim}{x}' \ , \quad j = 1,\ldots,N \ ,$$

which is solved on a computer. Fortunately the integrals in this system can be evaluated analytically, although for more complicated kernels numerical quadrature has to be used.

Once J_N^I has been calculated, we define J_N^{II} by the natural iteration

$$\hat{J}_N^{II}(\underset{\sim}{x}) = C_0 + \lambda \int_{D_0} k(\underset{\sim}{x},\underset{\sim}{x}')\hat{J}_N^I(\underset{\sim}{x}')d\underset{\sim}{x}' \ ,$$

or more explicitly

4.17 $$\hat{J}_N^{II}(\underset{\sim}{x}) = C_0 + \lambda \sum_{i=1}^{N} a_i \int_{\Delta_i} k(\underset{\sim}{x},\underset{\sim}{x}')d\underset{\sim}{x}' \ .$$

An error analysis for the collocation method for one-dimensional problems is given by Atkinson (see [2, pp.54-58]). The advantages of iteration of the collocation solution, again in one-dimensional cases, is described by Sloan, Noussair and Burn [18],[19]. Although a rigorous proof has not yet been given, it is expected that in two-dimensional cases, the iterated collocation solution will exhibit a higher order of convergence than the collocation solution.

The current distribution was calculated using both types of approximation, for the case when the cross section of the conductor was divided into 20 sub-rectangles (using 1 subdivision along the breadth and 20 along the length). The constant C_0 in (4.14) was chosen so that the total current flowing (according to the first approximation $\hat{J}_N^I$) was $\frac{b}{a}$. The following values in RMKS units for the parameters were used:

$$a = 0.1$$
$$b = 0.005$$
$$\mu = 4\pi \times 10^{-7}$$
$$\omega = 60$$
$$g = \frac{1}{2.83} \times 10^8$$

Since $\hat{J}_N^I$ is a piecewise constant function it can be easily tabulated. However, $\hat{J}_N^{II}$ is continuous over the cross-section so we evaluate it at 41 equally spaced points along the line joining the collocation points. These 41 points include the 20 collocation points and it is noticed that the solutions $\hat{J}_N^I$ and $\hat{J}_N^{II}$ have the same value at those points, as would be expected from equations (4.16) and (4.17). We tabulate the results only over the first 10 subrectangles; the distribution over the second 10 follows a similar pattern by symmetry.

	$\lvert \hat{J}_N^I \rvert$ (Collocation)	$\lvert \hat{J}_N^{II} \rvert$ (Iterated Collocation)
Edge of Conductor		1.2763
	1.2330	1.2330
		1.1949
	1.1595	1.1595
		1.1274
	1.0978	1.0978
		1.0708
	1.0460	1.0460
		1.0236
	1.0031	1.0031
		0.9848
	0.9683	0.9683
		0.9538
	0.9411	0.9411
		0.9302
	0.9210	0.9210
		0.9135
	0.9077	0.9077
		0.9036
	0.9011	0.9011
Centre of Conductor		0.9003

Distribution of current using a 20 × 1 discretisation of cross section.

The results tabulated are the approximate solutions of equation (4.13). Corresponding solutions to equation (4.12) (with $r_0 = a$) are obtained by the formulae :

$$J_N^I(ax,ay) = \hat{J}_N^I(x,y) \quad , \qquad (x,y) \in [0,1] \times [0,b/a]$$

for the collocation solution, and

$$J_N^{II}(ax,ay) = \hat{J}_N^{II}(x,y) \quad , \qquad (x,y) \in [0,1] \times [0,b/a]$$

for the iterated collocation solution.

ACKNOWLEDGEMENT

The author would like to thank Professor Ian Sloan for his help and encouragement.

REFERENCES

[1] Ahner, J.F. Integral equations and the interior Dirichlet potential problem, *J. Approx. Theory*, 22, 331-339, 1978.

[2] Atkinson, K.E. A survey of numerical methods for the solution of Fredholm integral equations of the second kind, *SIAM*, Philadelphia , Pa., 1976.

[3] Barnard, A.C.L., Duck, I.M. and Lynn, M.S. The application of electromagnetic theory to electrocardiology I, *Biophys. J.*, 7, 443-462, 1967.

[4] Barnard, A.C.L., Duck, I.M., Lynn, M.S. and Timlake, W.P. The application of electromagnetic theory to electrocardiology II, *Biophys. J.*, 7, 463-491, 1967.

[5] Burton, A.J. The solution of Helmholtz' equation in exterior domains using integral equations, Report NAC 30, National Physical Laboratory, Teddington, Mx., U.K., 1973.

[6] Fredholm, I. Sur une classe d'équations functionelles, *Acta. Math.*, 27, 365-390, 1903.

[7] Graham, I.G. and Sloan, I.H. On the compactness of certain integral operators, *J. Math. Anal. Appl.*, 68, 580-594, 1979.

[8] Jawson, J.A. and Symm, G.T. Integral Equations Methods in Potential Theory and Elastostatics, Academic Press, London, 1977.

[9] Kantorovich, L.V. and Akilov, G.P. Functional Analysis in Normed Spaces, Pergamon Press, Oxford, 1964.

[10] Kellogg, O.D. Foundations of Potential Theory, Dover, New York, 1953.

[11] Kleinman, R.E. and Roach, G.F. Boundary integral equations for the three-dimensional Helmholtz equation, *SIAM Review*, 16, 214-236, 1974.

[12] Lonseth, A.T. Sources and applications of integral equations, *SIAM Review*, 19, 241-278, 1977.

[13] Lynn, M.S. and Timlake, W.P. The numerical solution of singular integral equations of potential theory, *Numer. Math.* 11, 77-98, 1968.

[14] Noble, B. A bibliography on : Methods for solving integral equations. Author listing, M.R.C. Technical Summary Report #1176, Madison, Wisconsin, 1971.

[15] Noble, B. A bibliography on : Methods for solving integral equations. Subject listing, M.R.C. Technical Summary Report #1177, Madison, Wisconsin, 1971.

[16] Roach, G.F. On the approximate solution of elliptic self adjoint boundary value problems, *Arch. Ration. Mech. Anal.*, 27, 243-254, 1967.

[17] Silvester, P. Modern Electromagnetic Fields, Prentice-Hall, Englewood Cliffs, N.J., 1968.

[18] Sloan, I.H., Noussair, E. and Burn, B.J. Projection methods for equations of the second kind, *J. Math. Anal. Appl.* 69, 84-103, 1979.

[19] Sloan, I.H. A review of numerical methods for Fredholm equations of the second kind, this publication.

[20] Sternberg, W.J. and Smith, T.L. The Theory of Potential and Spherical Harmonics, University of Toronto, Toronto, Ontario, 1952.

[21] Taylor, A.E. Introduction to Functional Analysis, John Wiley, New York, N.Y., 1958.

[22] Tikhonov, A.N. and Samarskii, A.A. Equations of Mathematical Physics, Pergamon Press, Oxford, 1963.

[23] Ursell, F. On the exterior problems of acoustics, *Proc. Camb. Phil. Soc.*, 74, 117-125, 1973.

SUPERCONVERGENCE FOR SECOND KIND INTEGRAL EQUATIONS

G.A. Chandler
Computing Research Group/Pure Mathematics, SGS,
Australian National University, Canberra, A.C.T.

One of the latest buzz words in the numerical solution of differential equations is "superconvergence". The reason for the enthusiasm is quite clear; methods which give only an $O(h^{r+1})$ approximation to the solution overall were discovered to give $O(h^{2r})$ estimates of certain properties of the solution. One of the simplest examples was the solution of a second order ordinary differential equation by Galerkin's method using splines of degree r as basis functions. In general the error is $O(h^{r+1})$, but Douglas and Dupont [5] showed that the error is $O(h^{2r})$ at the knot points. It is obviously of interest to know about such things when interpreting the results of a numerical computation.

These higher order error estimates also occur in the numerical solution of integral equations. Indeed in one of the early papers on superconvergence de Boor and Swartz [2] start off by writing their differential equation $\sum_{i=0}^{m} a_i u^{(i)} = f$ as an integral equation $(I-K)(u^{(m)}) = f$.

This paper gives an overview of superconvergence phenomena for Galerkin approximate solutions of the linear Fredholm integral equation of the second kind;

$$u_0(x) - Ku_0(x) = f(x) , \qquad 0 \le x \le 1 , \tag{1}$$

where the operator K is defined by

$$(Ku)(x) = \int_0^1 k(x,\xi)u(\xi)d\xi \; .$$

The mathematical theory which underlies superconvergence results is developed in section 2. The three subsequent sections describe ways of using superconvergence to obtain more accurate approximations to the solution than given by the original Galerkin approximation. Section 1 contains the various mathematical preliminaries.

1. PRELIMINARIES

Throughout this paper attention is confined to methods which seek to approximate the solution, u_0, of (1) by a spline. It is not obvious that this can be done satisfactorily. For example, if the solution behaves as $x^{\frac{1}{2}}$ near the origin only modest accuracy is achieved. The purpose of this section is to put sufficient regularity assumptions on the kernel function so that u_0 is sufficiently "pleasant" for spline approximation to work well.

For the purposes of the mathematical analysis $I - K$ is regarded as a linear operator on $L^2([0,1])$. It is assumed that the problem (1) is *well posed* in the sense that for all $f \in L^2([0,1])$ there exists a unique solution $u_0 \in L^2([0,1])$ such that (1) holds.

For a given integer n let Δ_n (or just Δ if n is understood) denote the partition of $[0,1]$ given by the knot points $x_i = ih$, where $h = 1/n$ and $i = 0(1)n$. For any integers r and ν, with $0 \le r$ and $-1 \le \nu \le r$, define the polynomial splines of degree r and continuity ν, denoted $Sp_h(C^\nu, P_r)$ or S_h for short, to be the set of functions ϕ such that :

(i) ϕ is a polynomial of degree r on each of the subintervals $[x_{i-1}, x_i]$, and

(ii) if $\nu \ge 0$ then $\phi \in C^\nu$. (where C^ν denotes the set of functions with ν continuous derivatives)

The usual L^2 inner product and norm are denoted by $(\cdot,\cdot)$ and $\|\cdot\|$ respectively. For any integer k let $W^k(\Delta)$ denote the Sobolev space of functions u whose weak (distributional) k^{th} derivative exists between the knot points and for which the Sobolev semi-norm

$$|u|_{k,\Delta} = \left(\sum_{i=1}^{n} \int_{x_{i-1}}^{x_i} |u^{(k)}|^2 \right)^{\frac{1}{2}}$$

is finite. The notation W^k and $|u|_k$ are used when $n = 1$. The condition $u \in W^k$ is only a regularity requirement on the kth derivative. This is slightly more convenient for our purposes than the stronger requirement that the kth derivative be continuous between the knot points. For example if U_z is defined by

$$U_z(x) = \begin{cases} exp(x(z-1)) & 0 \le x \le z \\ exp(z(x-1)) & z \le x \le 1 \end{cases} ,$$

then $|U_z|_1 < \infty$, although U_z is not continuously differentiable. Note further that $|U_z|_{2,\Delta}$ is only finite when z is a knot point. Note also the useful fact that $u \in W^k$ implies $u \in C^{k-1}$.

The notation developed here is used to state the important approximation properties of splines.

<u>THEOREM 1</u> *For all functions* $u \in W^{r+1}(\Delta) \cap C^{\nu}$, *there exists a spline* $\phi \in Sp_h(C^{\nu}, P_r)$ *such that*

$$\|u-\phi\| \le Ch^{r+1}|u|_{r+1,\Delta} ,$$

where $C = C(r+1,\nu)$. ///

The notation $C = C(r+1,\nu)$ means that C is a number depending on $r + 1$ and ν.

A proof of the theorem can be found in de Boor and Fix [1].

To use this theorem to justify numerical methods the solution must be sufficiently smooth; that is $u_0 \in W^{r+1}(\Delta)$. Since the solution u_0 is not known, this smoothness can only be ensured if regularity assumptions are placed on K and f. Thus for a positive integer ℓ, K is ℓ-smoothing iff for all partitions Δ_n and integers i, $0 \le i \le \ell$, and all functions $u \in W^i(\Delta)$; $Ku \in W^{i+1}(\Delta)$ and

$$|Ku|_{i+1,\Delta} \le C|u|_{i,\Delta} , \tag{2}$$

where $C = C(\ell, K)$.

Assumptions of this form are essentially regularity assumptions on the kernel function of K. K is ℓ-smoothing for example if k is a Greens function of a nonsingular ordinary differential equation, or if k has $l+1^{th}$ derivatives.

Condition (2) is used in the following manner. If $f \in W^{l+1}(\Delta)$ then the equation $u_0 = f + Ku_0$ implies $u_0 = f + Kf + K^2f + \ldots + K^lf + K^{l+1}u_0$. Whence as $f, Kf, \ldots, K^lf \in W^{l+1}(\Delta)$ the implications $Ku \in W^1(\Delta), K(Ku) \in W^2(\Delta), \ldots, K^{l+1}u \in W^{l+1}(\Delta)$ show $u_0 \in W^{l+1}(\Delta)$. Thus knowing only the properties of k and f it can be shown that the unknown solution is sufficiently smooth to be approximated by a spline. This argument breaks down when the kernel is $k(x,\xi) = |x-\xi|^{-\frac{1}{2}}$. If $f = 1$ for example, $(K1)(x) = 2x^{\frac{1}{2}} + 2(1-x)^{\frac{1}{2}} \notin W^1$. In fact the actual solution will look like $x^{\frac{1}{2}}$ near the origin, and any spline approximation will be at most $O(h)$ (although this does not follow from theorem 1). The weak singularity of the kernel creates a genuine numerical difficulty even for a smooth f. The methods discussed here do not cope with this singularity. The strong regularity assumptions are required for the method to work well; not just for the proofs to go through.

2. Galerkin Methods

Given that there exists spline(s) close to the unknown solution of (1), the problem is how to construct one. The *Galerkin Method* provides one possibility.

Let $P_h : L^2([0,1]) \to S_h$ denote the orthogonal projection of $L^2([0,1])$ onto some spline subspace S_h. The *Galerkin approximation* $u_h \in S_h$ is defined by

$$\left(I - P_hK\right)u_h = P_hf \; . \tag{3}$$

The solution of equation (3) may be found computationally. Selecting a basis $\{\phi_i\}$ for S_h and writing $u_h = \Sigma\alpha_i\phi_i$, the vector $\alpha = (\alpha_i)^T$ is the solution of the matrix equation

$$A\alpha = \beta \quad ; \tag{4}$$

where $A_{ij} = \left((I-K)\phi_j, \phi_i\right)$ and $\beta_i = \left(f, \phi_i\right)$. The standard theory (Krasnoselskii *et. al.* [8]) shows that, for h sufficiently small, u_h is uniquely defined by (3) and that

$$\|u_0 - u_h\| \leq C \; inf \; \{\|u_0 - \phi\| : \phi \in S_h\} \quad .$$

Using theorem 1 to bound the right hand side proves the following theorem.

<u>THEOREM 2</u>. *Provided* K *is* r*-smoothing and* $f \in W^{r+1}$

$$\|u_0 - u_h\| \le Ch^{r+1}|f|_{r+1}$$

where $C = C(K,r,\nu)$. ///

However suppose we wish to find the value of the inner product (Φ, u_0) , where Φ is some suitably smooth function. What error is committed in using the obvious approximation (Φ, u_h) ? Clearly

$$|(\Phi,u_0) - (\Phi,u_h)| = |(\Phi,u_0-u_h)| \le \|\Phi\| \ \|u_0-u_h\| ,$$

which shows the error is $O(h^{r+1})$ (by theorem 2).

In fact some trickery shows that the order of convergence is twice as high, and an $O(h^{2r+2})$ error bound is possible. It is this doubling of the order of convergence that leads to the term *superconvergence*.

<u>THEOREM 3</u>. *If* K *and* K^* *are* r*-smoothing and* $f \in W^{r+1}$ *, then for all* $\Phi \in W^{r+1}(\Delta) \cap C^\nu$

$$|(\Phi, u_h - u_0)| \le Ch^{2r+2}|\Phi|_{r+1,\Delta}|f|_{r+1}$$

where $C = C(K,r,\nu)$.

Proof : Since $I - K$ is well posed the Fredholm alternative shows $I - K^*$ is also well posed. Thus there exists a unique $\Psi \in L^2$ such that $(I-K^*)\Psi = \Phi$. Furthermore as K^* is r-smoothing $\Psi \in W^{r+1}(\Delta) \cap C^\nu$ and $|\Psi|_{r+1,\Delta} \le C|\Phi|_{r+1,\Delta}$.

For any $\phi \in S_h$

$$\begin{aligned}(\Phi, u-u_h) &= ((I-K^*)\Psi, u_0-u_0) \\ &= (\Psi, (I-K)(u_0-u_h)) \\ &= (\Psi-\phi, (I-K)(u_0-u_h)) .\end{aligned}$$

(Because $(\phi, (I-K)(u_h-u_0)) = 0$ by (3)).

Thus by Theorem 1

$$|(\Phi,u-u_h)| \le C\|\Psi-\phi\| \; \|(I-K)(u_0-u_h)\|$$
$$\le Ch^{2r+2}|\Phi|_{r+1,\Delta}|f|_{r+1} \; . \qquad ///$$

Theorem 3 may be modified when $\Phi \in W^m(\Delta) \cap C^\nu$ for $\nu < m < r+1$. Then the error bound becomes $Ch^{r+1+m}|\Phi|_{m,\Delta}|f|_{r+1}$. In the finite element literature theorem 3 is known as a negative norm estimate.

The intuitive reason for this result is that the error $u_0 - u_h$ is highly oscillatory. When calculating (Φ,u_h-u_0) the positive and negative errors cancel systematically, leaving only a small $O(h^{2r+2})$ error.

Sometimes when solving integral equations it is necessary to produce approximations to (Φ,u_0) ; for example u_0 may be the solution to a boundary integral equation and values of the solution to the original problem in the interior are required. However the viewpoint taken here is that the ultimate aim of any numerical procedure is to obtain accurate approximations to the solution at all points. From this perspective Theorem 3 is unsatisfactory. It would be preferable to have superconvergence of an estimate of the actual solution. More practically stated, we would like to use the superconvergence theory to produce methods which remove the highly oscillatory error in u_h and produce a higher order approximation to the solution.

3. The Natural Iteration

Presently the best means of obtaining a superconvergent estimate of the solution is the natural iteration. Here the Galerkin solution u_h is used to calculate the natural iterate

$$u_h^* = f + Ku_h \; . \tag{5}$$

Since $K\phi_i$ has been calculated originally to set up the Galerkin equation (4), Ku_h can be calculated cheaply. For Galerkin methods this idea is due to Prof. Sloan. More information and some numerical examples can be found in his paper in these proceedings and the references cited there.

The basis for the numerical analysis of u_h^* is the observation (due to Sloan) that

$$u_h^* - u_0 = (f+Ku_h) - (f+Ku_0) = K(u_h-u_0) .$$

Hence

$$(u_h^*-u_0)(x) = (k_x, u_h-u_0) , \tag{6}$$

where k_x is the function defined by $k_x(\xi) = k(x,\xi)$. Equation (6) is in the right form for the immediate application of Theorem 3, provided of course $k_x \in W^{r+1}(\Delta) \cap C^\nu$. This will be illustrated by two examples.

Firstly consider the very smooth kernel

$$k^{(1)}(x,\xi) = e^{x-\xi} .$$

(Note in passing that as $k^{(1)}$ is so smooth a method using Chebyshev polynomials as basis functions will give a geometric rate of convergence. Splines are not the ideal method). Here $|k_x^{(1)}|_{r+1,\Delta} \le 1$ for all x, and it follows by using Theorem 3 in equation (6) that

$$\|u_h^*-u_0\|_\infty \le Ch^{2r+2}|f|_{r+1} .$$

($\|u\|_\infty$ denotes the uniform norm $\|u\|_\infty = sup\ \{|u(x)| : x \in [0,1]\}$.) Thus the natural iteration doubles the order of convergence, and u_h^* superconverges to u_0.

A slightly more difficult example is the kernel

$$k^{(2)}(x,\xi) = \begin{cases} (e^\xi-1)(e^{x-1}-1) \ / \ (e-1)e^{\xi-1} & \xi \ge x \\ (e^x-1)(e^{\xi-1}-1) \ / \ (e-1)e^{\xi-1} & \xi \ge x \ . \end{cases}$$

($k^{(2)}$ is the Greens function of the ordinary differential operator $D^2u - Du$, $u(0) = u(1) = 0$ If x is a knot point, $x = x_i$, then $k_{x_i} \in W^{r+1}(\Delta) \cap C^0$. Thus provided the basis splines are chosen with $\nu = -1$ or $\nu = 0$,

$$|(u_h^*-u_0)(x_i)| \le Ch^{2r+2}|k_{x_i}|_{r+1,\Delta}|f|_{r+1} .$$

Whence noting that $|k_{x_i}|_{r+1,\Delta}$ can be bounded independently of the particular knot point chosen,

$$|(u_h^*-u_0)(x_i)| \le Ch^{2r+2}|f|_{r+1}$$

where $C = C(K,r,\nu)$.

When x is not a knot point the natural iteration does not lead to the full superconvergence. The modification described after the proof of Theorem 3 gives $|(u_h^*-u_0)(x)| \le O(h^{r+2})$. More careful analysis shows that k_x can be approximated to within $O(h^{3/2})$, and thus $h^{r+5/2}$ is possible. But ultimately an examination of the singular values of K shows the ratio

$$\gamma_h(u) = \frac{\inf \{\|K(u-\phi)\| : \phi \in S_h\}}{\inf \{\|u-\phi\| : \phi \in S_h\}}$$

is bounded below in the sense that

$$\inf \{\gamma_h(u) : u \in L^2\} \ge Ch^2$$

for some $C = C(K,r,\nu)$. Thus there is a definite limit to the order of improvement that can be proved globally using the straight-forward superconvergence arguments.

A modified form of the natural iteration does restore the full superconvergence. Define the function r_x by $(I-K^*)r_x = k_x$. Then for each x there is a sequence of constants $a_0(x),\ldots,a_r(x)$ such that

$$r_x = S_x + R_x ;$$

where

$$S_x(\xi) = \sum_{i=0}^{r} a_i(x)(x-\xi)_+^i$$

and

$$R_x \in W^{r+1} .$$

Now instead of u_h^* calculate

$$u_h^{**} = f + Ku_h + (S_x, f-(I-K)u_h) .$$

Then the argument used in the proof of theorem 3 shows

$$|(u_h^{**}-u_0)(x)| = |(R_x,(I-K)(u_0-u_h)| \le Ch^{2r+2}|f|_{r+1}$$

This idea appears in Dupont [6] for ordinary differential equations. Suitably modified it can be applied to any kernel which is smooth except for a discontinuity (as opposed to a singularity) along the diagonal $x = \xi$.

The notable feature about these methods is that it is possible to obtain $O(h^{2r+2})$ convergence; but make assumptions only about the $r+1^{th}$ derivatives. This is an advantage of the iterated Galerkin method over most other methods. To prove the same order of convergence for the Nyström method requires the existence of the $2r+2^{th}$ derivatives of both the solution and the kernel. The product integration method requires the $2r+2^{th}$ derivative of the solution and the $r+1^{th}$ derivative of k . This advantage of Galerkin methods has not been explored; although it must be balanced against the disadvantage of having to calculate the integrals in equation (4).

More details, including some explicit theorems are given in Chandler [7].

4. Special Points

Superconvergence is most frequently associated with the location of special points at which a higher order of convergence takes place. As stated in the beginning, for differential equations the order of convergence at knot points is almost double the order of convergence obtained globally. ($O(h^{2r})$ compared with $O(h^{r+1})$) Special points can also be found for integral equations but they are not the knot points : rather they are the Gauss-Legendre points on each sub-interval. Furthermore the order of convergence increases only from $O(h^{r+1})$ to $O(h^{r+2})$. Richter [9] shows when $Sp_h(C^{-1},P_r)$ or $Sp_h(C^0,P_{2\rho+1})$ (i.e. $r = 2\rho+1$ is odd) are used as a basis, that

$$|(u_h-u_0)(\xi_{ij})| \leq Ch^{r+2}|u_0|_{r+2} \tag{7}$$

where $\xi_{i0},\dots\xi_{ir}$ are the Gauss-Legendre points on the ith interval. The proof proceeds by observing that $P_h u_h^* = u_h$. Hence the arguments of section 3 show

$$\|P_h u_0 - u_h\|_\infty = \|P_h(u_0-u_h^*)\|_\infty \leq Ch^{r+2}|u_0|_{r+2}$$

for $C = C(K,r,\nu)$; provided $k_x \in W^1$ for each x . (We have also used the fact (de Boor [3]) that P_h is bounded as a map $L^\infty \to L^\infty$) Whence

$$|(u_h-u_0)(\xi_{ij})| = |(P_hu_0-u_0)(\xi_{ij})| + O(h^{r+2}) .$$

Thus to prove (7) it is only necessary to look at the form of the error in least squares approximation (i.e. $P_hu_0 - u_0$) : the details are in Richter.

The disadvantages of this approach are the higher regularity conditions imposed on u_0 , and more importantly that $O(h^{r+2})$ is the highest error estimate obtainable. (Looking at the $Sp_h(C^{-1},P_r)$ least squares approximation to x^{r+2} shows that $O(h^{r+3})$ cannot be obtained in (7)).

However, suppose the projection P_h in equation 3 is the interpolatory projection onto $Sp_h(C^{-1},P_r)$ defined by

$$(P_hu)(\xi_{ij}) = u(\xi_{ij}) .$$

Then u_h is the collocation approximation to u_0 , and u_h^* is the product integration approximation to u_0 (See Prof. Sloan's article in this proceedings). Provided k has $r+1^{th}$ derivatives and $u_0 \in W^{2r+2}$, $|u_h^*-u_0| \le O(h^{2r+2})$ and $P_hu_h^* = u_h$ imply

$$|(u_h-u_0)(\xi_{ij})| \le Ch^{2r+2}|u_0|_{2r+2} .$$

Thus it is natural to find special points for product integration/collocation; but not for Galerkin's method.

On the other hand the analogue of Theorem 3 for product integration is less natural and requires strong regularity conditions.

5. SUPERINTERPOLATION

This last section describes another method of obtaining a superconvergent estimate of the actual solution. Its inclusion is somewhat tentative. The method is probably more useful for differential equations. Nevertheless it does have some advantages when applied to integral equations. It provides a high order spline approximation to the solution and would be useful if an approximation in this form were required. Furthermore this high order spline is constructed from the Galerkin solution, and thus would be an alternative if the evaluations of Ku_h required by the natural iteration were expensive.

As this is a preliminary description attention will be confined initially to the case $S_h = Sp_h(C^{-1}, P_1)$. For motivation suppose the kernel function $k = 0$. Then the Galerkin approximation is the familiar L^2 least squares approximation to u_0 . (Equation (3) reduces simply to the condition $u_h = P_h u_0$) Theorem 1 immediately gives

$$\|u_h - u_0\| \le Ch^2 |u_0|_2 \ .$$

Using a well known result about splines (immediately from Schultz [10] p.31) shows $u_h^{(-2)}$ ($u^{(-m)}$ denotes the m-fold indefinite integral of u) is the Hermite cubic spline (i.e. $u_h^{(-2)} \in Sp_h(C^1, P_3)$) interpolating $u_0^{(-2)}$. That is for all i

$$u_h^{(-2)}(x_i) = u_0^{(-2)}(x_i) \ , \tag{8.1}$$

and

$$u_h^{(-1)}(x_i) = u_h^{(-1)}(x_i) \ . \tag{8.2}$$

Hence knowing just u_h allows us to calculate $u_0^{(-1)}$ and $u_0^{(-2)}$ exactly at the knot points. To estimate u_0 itself simply apply a high order second difference operator to $u^{(-2)}$. In particular define

$$(\Delta_h^2 u)(x) = \frac{1}{h^2}\{2u(x-h) + \frac{h}{2}u^{(1)}(x-h) - 4u(x) - \frac{h}{2}u^{(1)}(x+h) + 2u(x+h)\}$$

and

$$(\Delta_h^3 u)(x) = \frac{1}{h^3}\{-\frac{15}{2}u(x-h) - \frac{3h}{2}u^{(1)}(x-h) - 12hu^{(1)}(x) - \frac{3h}{2}u^{(1)}(x+h) + \frac{15}{2}u(x+h)\} \ ,$$

and observe that

$$|(\Delta_h^2 u)(x) - u^{(2)}(x)| \le Ch^4 |u|_6$$

and

$$|(\Delta_h^3 u)(x) - u^{(3)}(x)| \le Ch^4 |u|_7 \ .$$

(Because Δ_h^2 (resp. Δ_h^3) gives the exact values of $u^{(2)}$ (resp. $u^{(3)}$) whenever u is a polynomial of degree ≤ 5 (resp. 6)). Therefore at the interior knot points

$$\tilde{\alpha}_i = (\Delta_h^2 u_h^{(-2)})(x_i)$$

and

$$\tilde{\beta}_i = (\Delta_h^3 u_h^{(-2)})(x_i)$$

are $O(h^4)$ approximations to

$$\alpha_i = u_0(x_i)$$

and

$$\beta_i = u_0^{(1)}(x_i)$$

respectively. At the end point $x_i = 0$ (resp. $x_i = 1$) analogous forward (backward) difference schemes involving the points $0,h$ and $2h$ ($1,1-h$ and $1-2h$) can be used. It is a simple matter to compute the Hermite cubic spline U_h such that $U_h(x_i) = \tilde{\alpha}_i$ and $U_h^{(1)}(x_i) = \tilde{\beta}_i$. It now follows from the interpolating properties of cubic Hermites that

$$\|U_h - u_0\| \leq Ch^4 |u_0|_4 \quad .$$

The important observation is that $\tilde{\alpha}_i$ and $\tilde{\beta}_i$ and thus U_h are constructed directly from u_h . Thus if we return to the non-trivial case in which u_h is the Galerkin approximation to the integral equation (1) with $k \neq 0$ it is still possible to construct U_h as outlined above. Now however it is not possible to use a theorem about splines to show (8.1) and (8.2) and hence that $\tilde{\alpha}_i$ and $\tilde{\beta}_i$ approximate α_i and β_i . Instead the super-convergence theory of section 2 is used.

THEOREM 4. *Suppose* $k \in W^2([0,1] \times [0,1])$ *and* $u_0 \in W^4$. *Then*

$$|\alpha_i - \tilde{\alpha}_i| \leq Ch^4 |u_0|_4 \quad , \tag{9.1}$$

$$|\beta_i - \tilde{\beta}_i| \leq Ch^3 |u_0|_4 \tag{9.2}$$

and hence

$$\|U_h - u_0\| \leq Ch^4 |u_0|_4 \quad , \tag{10}$$

where $C = C(K)$.

Proof : By the triangle inequality

$$|\alpha_i-\tilde{\alpha}_i| = |u_0(x_i) - \Delta_h^2 u_h^{(-2)}(x_i)| \le |(u_0-\Delta_h^2 u_0^{(-2)})(x_i)| + |\Delta_h^2 (u_0-u_h)^{(-2)}(x_i)| \; . \tag{11}$$

Because Δ_h^2 is a high order difference scheme,

$$|(u_0-\Delta_h^2 u_0^{(-2)})(x_i)| \le \begin{cases} Ch^4 |u_0|_4 \\ Ch^2 |u_0|_2 \; . \end{cases} \tag{12}$$

Thus the first term of (11) is $O(h^4)$. The trick in dealing with the second is the well known fact that for any function u

$$(\Delta_h^2 u^{(-2)})(x) = (\Phi_x^h, u) \; ,$$

where $\Phi_x^h(\xi) = \Phi^h(\xi - x)$ and Φ^h is defined by

$$\Phi^h(\xi) = \begin{cases} \frac{3}{2h} + \frac{2}{h^2}\xi & -h \le \xi < 0 \\ \frac{3}{2h} - \frac{2}{h^2}\xi & 0 \le \xi \le h \\ 0 & \text{otherwise.} \end{cases}$$

Therefore

$$\begin{aligned} \Delta_h^2 (u_0-u_h)^{(-2)}(x_i) &= (\Phi_x^h, u_0-u_h) \\ &= ((I-K^*)\Phi_{x_i}^h, u_0-u_h) + (K^*\Phi_{x_i}^h, u_0-h_0) \\ &= (\Phi_{x_i}^h, (I-K)(u_0-u_h)) + (K^*\Phi_{x_i}^h, u_0-u_h) \; . \end{aligned} \tag{13}$$

The first term in (13) is zero because $\Phi_{x_i}^h \in Sp_h(C^{-1}, P_1)$ and the orthogonality properties of the Galerkin approximation. (In fact this is just the proof of (8.1)) The second term of (13) is bounded by using the second half of (12) to show

$$|(\Phi_{x_i}^h, K(u_0-u_h)) - K(u_0-u_h)(x_i)| \le Ch^2 |K(u_0-u_h)|_2 \; .$$

The methods of section 3 show that $\|K(u_0-u_h)\| \le Ch^4 |u_0|_2$, and by the condition placed on k in the statement of the theorem

$$|K(u_0-u_h)|_2 \le C\|u_0-u_h\| \le Ch^2|u_0|_2 \ .$$

This proves (9.1) for the interior knot points.

The proof of (9.1) at the end points $x_i = 0$ and $x_i = 1$ is similar, as is the proof of (9.2). Finally given the fact (Schultz [10] p.34) that the cubic Hermite spline interpolating α_i and β_i has error $\le Ch^4|u_0|_4$, (10) follows from (9.1) and (9.2) by elementary means. ///

Theorem 4 should generalize without trouble to higher order splines $Sp_h(C^{-1},P_r)$. Alternatively the process can be modified to deal with $Sp_h(C^0,P_1)$. Define the high order difference approximation

$$(\tilde{\Delta}_h^2 u)(x) = \frac{1}{h^2}\{-\tfrac{1}{6}u(x-2h) + \tfrac{8}{3}u(x-h) - 5u(x) + \tfrac{8}{3}u(x+h) - \tfrac{1}{6}u(x+2h)\} \ .$$

Then $V_h = (\tilde{\Delta}_h u_h^{(-2)}) \in Sp_h(C^2,P_3)$ and has $O(h^4)$ accuracy. There is a function Ψ^h analogous to Φ^h; and it is easily seen that $V_h = \Psi^h * u_h$. Thus V_h is in fact obtained by smoothing u_h; which is clearly a way to eliminate the oscillatory portion of the error in u_h. In the finite element literature V_h appears as the averaging method of Bramble and Schatz [4]. In fact V_h can also be constructed in the $Sp(C^{-1},P_1)$ case; but because of the large support of $\tilde{\Delta}_h^2$ compared with Δ_h^2, V_h will be less accurate than U_h. ($|(\Delta_h^2 u^{(-2)}-u)(x)| \,/\, |(\tilde{\Delta}_h^2 u^{(-2)}-u)(x)|$ is approximately 1/8).

Acknowledgement

The facts about weakly singular kernels described at the end of section 1 were explained to me by Ivan Graham. I have had valuable discussions with Alex McNabb about $k^{(2)}$.

I thank Bob Anderssen, Frank de Hoog and Mark Lukas for the dry run.

REFERENCES

[1] C. de Boor and G.J. Fix; "Spline approximation by quasi-interpolants", *J. Approx. Th.* 8 (1973), p.19-45.

[2] C. de Boor and B.K. Swartz; "Collocation at Gaussian points", *SIAM J. Numer. Anal.* 10 (1973), p.582-606.

[3] C. de Boor; "A bound on the L^∞ norm of L^2 approximation by splines in terms of a global mesh ration", *Maths of Comp.* 30 (1976), p.765-771.

[4] J.H. Bramble and A.H. Schatz; "High order local accuracy by averaging in the finite element method", *Maths. Comp.* 31 (1977), p.94-111.

[5] J. Douglas Jr. and T. Dupont; "Galerkin approximations for the two point boundary value problem using piecewise polynomial spaces", *Numer. Math.* 22 (1974).

[6] T. Dupont; "A unified theory of superconvergence for Galerkin methods for two point boundary value problems", *SIAM J. Numer. Anal.* 13 (1976).

[7] G.A. Chandler; "Global superconvergence of iterated Galerkin solutions for second kind integral equations", submitted for publication.

[8] M.A. Krasnoselskii et. al.; *Approximate Solution of Operator Equations*, Wolters-Noordhoff, 1972. (Original Russian edition 1969).

[9] G.R. Richter; "Superconvergence for piece-wise polynomial Galerkin approximations for Fredholm integral equations of the second kind", Numer. Math. 31 (1978), p.63-70.

[10] M.H. Schultz; *Spline Analysis*, Prentice-Hall, 1973.

REVIEW OF FREDHOLM EQUATIONS OF THE FIRST KIND

F.R. de Hoog
Division of Mathematics and Statistics, CSIRO, Canberra, A.C.T.

1. INTRODUCTION

Fredholm integral equations of the first kind

(1.1) $$\int_a^b k(t,s)f(s)ds = g(t) \ , \qquad a \le t \le b$$

arise regularly in applications. The form of the kernel k and the data g depend heavily on the nature of the problem. It is therefore important to examine such equations in the context of the associated application. Various examples are discussed in §2.

Numerically, equations of the form (1.1) are generally very difficult to solve. To understand this intuitively, consider the case when $k(t,s)$ is absolutely continuous as a function of s. Then, from the Riemann-Lebesgue Theorem,

$$\int_a^b k(t,s)\cos(ns)ds \to 0 \qquad \text{as} \quad n \to \infty \ .$$

This means that an arbitrarily small perturbation in g can give rise to a perturbation of order one in the solution f. If we now approximate (1.1) by a consistent numerical scheme we should expect that the scheme will mimic this behaviour and will therefore be badly conditioned.

An illustration of another difficulty is provided by the case when $k(t,s) = \cos(t-s)$. Then, g must have the form

$$g = c_1 \cos t + c_2 \sin t$$

if a solution exists (i.e. the left and right hand sides of the equation must be compatible) and

$$\int_a^b f(s)\cos(s)\,ds = c_1 , \qquad \int_a^b f(s)\sin(s)\,ds = c_2 .$$

Clearly, the solution is not unique. It is important to note that even if (1.1) is consistent, it is likely that an arbitrarily chosen numerical scheme for its solution will fail to be consistent.

The above concepts will be made more precise in the sequel and have only been introduced here to demonstrate that there are fundamental differences between first and second kind Fredholm equations. In the light of this, it is not surprising that an approach different to that used for the numerical solution of second kind equations is needed.

2. Applications

In order to appreciate some of the difficulties which are discussed below, it is necessary to examine some representative applications.

The data recorded when a phenomenon is measured has often been smoothed to some degree by the instrument used. Thus, a fairly common problem is to reconstruct from a measurement g , the actual signal f . Typically, f and g are related by an equation of the form

$$\int_{-\infty}^{+\infty}\int_{-\infty}^{+\infty} k(x-\xi,y-\eta)f(\xi,\eta)\,d\xi d\eta = g(x,y) . \tag{2.1}$$

The fact that the kernel is of convolution type means that the smoothing is invariant with respect to the variables x and y . Although (2.1) is two dimensional, most of the theory applicable to (1.1) generalizes to higher dimensions. An example of (2.1) arises when the kernel is the Gaussian point spread function

$$k(\xi,\eta) = \frac{1}{2\pi\sigma^2} \exp\left[-(\xi^2+\eta^2)/2\sigma^2\right]$$

which is the prototype for random media degradations [13], flash radiography [7] and X ray pictures [1]. A similar kernel is appropriate when restoring turbulent degraded images [6].

From meteorology [12] we have the following problem. The distribution of the intensity of diffracted light (this is the data g) is related to the distribution of the radii of spherical particles in suspension (this is the unknown f) by a first kind Fredholm equation with the kernel

$$k(t,s) = [\sin(ts)/ts - \cos(ts)]^2 .$$

This problem is also encountered in polymer chemistry.

Another problem from polymer chemistry is the calculation of molecular weight distributions from the total weight concentrations that can be measured in sedimentation experiments [11]. The equation in this case is

$$\frac{1}{\lambda}\int_0^\infty \frac{se^{-s\xi}}{1-e^{-s}}\, f\!\left(\frac{s}{\lambda}\right)dx = g(t) \ , \qquad a \le t \le b$$

where $\xi = (b^2-t^2)/(b^2-a^2)$. Note that in this case the integration is over $(0,\infty)$ while the data is measured only on (a,b) .

In the applications considered so far, the data has been obtained from measurement and will therefore contain noise. However, in some cases the data is given analytically and the following is an example. Let D be a two dimensional region bounded by a closed curve C and consider the Dirichlet problem

$$\Delta u = 0 \qquad \text{in } D$$

$$u = g \qquad \text{in } C \ .$$

The solution may be represented as a single layer potential

$$u(P) = -\int_C \ln[\rho(P,P(s))]f(s)ds \tag{2.2}$$

where s is the arc length and $\rho(P,P(s))$ is the distance between the points $P \in D$ and $P(s) \in C$. If we now consider $P \to P(t) \in C$, we obtain the following equation for f

$$-\int_C \ln[\rho(P(t),P(s))]f(s)ds = g(P(t)) \ . \tag{2.3}$$

If an approximation to f can be found, it can then be used in (2.2) to obtain an approximation to u in D . This and related problems are discussed at some length by L. Wardle in these proceedings. We would however like to stress that, when solving (2.2) numerically, we are not generally interested in convergence to the single layer potential f , but are really interested in convergence to u . This simple fact is often missed in the literature.

3. Ill-Posed Nature of First Kind Equations

Equation (1.1) may be written in operator notation as

$$Kf = g \tag{3.1}$$

where $K : X \to Y$ and X and Y are complete metric spaces. According to Hadamard, (3.1) is a well posed problem if ;

(a) the solution $f \in X$ exists and is unique in X for every $g \in Y$

(b) the solution $f \in X$ depends continuously on g .

A problem that is not well posed is called ill-posed or improperly posed.

Of course, when talking about improperly posedness, it is necessary to take into consideration the metric spaces X and Y with which we are working. Indeed, for most 'ill-posed problems' it is possible to choose these spaces such that the problem becomes well posed (see for example Ivanov [8]). This is of little practical use however since the application in which the problem occurs dictates the spaces that are appropriate. In practice $L_2(a,b)$ and $C[a,b]$ are usually appropriate.

To examine the question of ill-posedness further, let us consider the case when $X = Y = L_2(a,b)$ and

$$\int_a^b \int_a^b |k(t,s)|^2 dtds < \infty .$$

Then, the operator K is compact and it follows that the problem must be ill-posed as the inverse of K (if it exists) is unbounded. The degree of 'ill-posedness' of the problem can be examined by considering the singular value decomposition (see for example Smithies [16])

$$k(t,s) = \sum_{r=0}^{\infty} \lambda_r \phi_r(t) \psi_r(s) .$$

Here, ϕ_r and ψ_r , $r = 0,1,\ldots$ are orthonormal singular functions of K which satisfy

$$K\psi_r = \lambda_r \phi_r \quad \text{and} \quad K^* \phi_r = \lambda_r \psi_r$$

or equivalently

$$KK^* \phi_r = \lambda_r^2 \phi_r \quad \text{and} \quad K^* K \psi_r = \lambda_r^2 \psi_r$$

where $\lambda_0^2 > \lambda_1^2 > \ldots$. Note that when the kernel is symmetric (i.e. $k(t,s) = k(s,t)$) , the singular functions are the eigenfunctions. Furthermore, $\{\psi_r\}$ and $\{\phi_r\}$ form a basis for $L_2(a,b)$.

If g belongs to the range of K , then $\lambda_r = 0$ implies that $(g,\phi_r) = 0$ and we may write the solution to our problem as

$$f = \sum_{r=0}^{\infty} (g,\phi_r)\psi_r/\lambda_r . \tag{3.2}$$

In order that this solution belong to $L_2(a,b)$, we require

$$\|f\|^2 = \sum_{r=0}^{\infty} [(g,\phi_r)/\lambda_r]^2 < \infty \ .$$

Furthermore, this solution will be unique if and only if $\lambda_r \neq 0$, $r = 0,1,\ldots$. From Hilbert-Schmidt theory, the sequence of singular values $\{\lambda_r\}$ has zero as its only possible limit point. Thus either, $\lambda_k = 0$, $k \geq k_0$ or zero is a limit point. In the former case, the problem does not have a unique solution while in the latter case, a perturbation of ϕ_r in the right hand side of (1.1) leads to a perturbation ψ_r/λ_r (which will become arbitrarily large as $r \to \infty$) in the solution. Thus, for the case when a unique solution exists, (i.e. $\lambda_r \neq 0$) a measure of the 'conditioning' of the problem is provided by the rate at which the singular values go to zero. For symmetric kernels, this question has been investigated by Weyl [22].

<u>THEOREM 3.1</u> *If* $k(t,s) = k(s,t)$ *and* $\frac{\partial^q k(t,s)}{\partial t^q}$ $q = 0,\ldots,p$ *exist and are continuous, then*

$$|\lambda_r| = 0(r^{-p-\frac{1}{2}}) \ .$$

Smithies [17] has given a more precise result on the singular values.

<u>THEOREM 3.2</u> *Let* β *be such that* $1 < \beta \leq 2$. *If for some* $p \geq 0$, $k(t,s)$ *satisfies*

(a) $k^{(q)}(t,s) = \frac{\partial^q k(t,s)}{\partial t^q}$ *exists and is absolutely continuous in* t *for almost all* s, $q = 0,\ldots,r-1$

(b) $k^{(p)}(t,s) \in L_\beta$ *for almost all* s

(c) $\int_a^b \left\{\int_a^b |k^{(p)}(t+\theta,s) - k^{(p)}(t-\theta,s)|^\beta dt\right\}^{2/\beta} ds \leq A|\theta|^{2\alpha}$ *for some* A *and all sufficiently small* θ, *and*

(d) *either* $p > 0$, $\alpha > 0$ *or* $p = 0$, $\alpha > 1/\beta - 1/2$ *then*

$$\lambda_r = 0(r^{-\alpha-1-p+1/\beta}) \ .$$

(Note that $k(t,s)$ is defined outside the range $a \leq t \leq b$ by its even extension if p is even and by its odd extension if p is odd.)

Basically, theorems 3.1 and 3.2 say that the smoother the kernel the worse will be the conditioning of the problem. For example, the kernels $\ln\left|\cos\left(\frac{t-s}{2}\right)\right|$ with $a = 0$, $b = 2\pi$ and $\exp(-|t-s|)$ with $a = 0$, $b = 1$ which are not very smooth have singular values which go to zero like $1/r$ and $1/r^2$ respectively. This conditioning is not very severe and is comparable to differentiating g once or twice. On the other hand, kernels that are analytic may have singular value that go to zero exponentially. This is the case for $k(t,s) = (1-\gamma^2)/(1+\gamma^2-2\gamma\cos[2\pi(t+s)])$ with $a = 0$, $b = 1$ which has singular values 1, $\pm\gamma^r$, $r = 1,2,\ldots$. Note however that when γ is close to 1, the kernel is peaked and the singular values decrease quite slowly initially. For example, if $\gamma = \cdot 9$, then $\gamma^{40} \sim \cdot 015$ and from a practical point of view the problem may be considered to be reasonably well conditioned.

4. Direct Discretization of First Kind Equations

It is of interest to consider some direct schemes for obtaining a numerical approximation to (1.1). To illustrate some of the difficulties associated with such schemes, consider the discretization

$$\Delta s \sum_{j=1}^{m} k(t_r,s_j)\omega_{j,m} f_j = g_r , \qquad r = 1,\ldots,n \tag{4.1}$$

where $t_r = a + (r-1)\Delta t$; $\Delta t = (b-a)/(n-1)$

and $s_j = a + (j-1)\Delta s$; $\Delta s = (b-a)/(m-1)$.

In matrix notation, we have

$$\Delta s\ KWf = g$$

and this equation is to be solved directly if $m = n$ or in the least squares sense if $n > m$. Hence, if K^TK is nonsingular (and this is not necessarily the case),

$$W\underset{\sim}{f} = \frac{1}{\Delta s}(K^TK)^{-1}K^T\underset{\sim}{g} . \tag{4.2}$$

Consider now $\underset{\sim}{f}_T$ and $\underset{\sim}{f}_S$ which are the approximate solutions obtained using the trapezoidal (i.e. $W_T = \mathrm{diag}(\frac{1}{2},1,1,\ldots,1,\frac{1}{2})$) and Simpson (i.e. $W_S = \mathrm{diag}(\frac{1}{3},\frac{4}{3},\frac{2}{3},\ldots,\frac{2}{3},\frac{4}{3},\frac{1}{3})$) quadrature weights respectively. Then, from (4.2)

$$W_T\underset{\sim}{f}_T = W_S\underset{\sim}{f}_S$$

and hence

$$\|\underset{\sim}{f}_T-\underset{\sim}{f}_S\|_2/\|\underset{\sim}{f}_S\|_2 = 1/3 .$$

Thus, at least one of the numerical schemes will not converge. Note however that the result is independent (at least superficially) of the conditioning of the problem.

From Theorems 3.1 and 3.2, the smoother the kernel, the worse is the conditioning of the problem and intuitively, we expect a similar result to hold for (4.1). To show that this is indeed the case, let $k(t,s)$ be p times continuously differentiable with respect to s. Then, there exist two distinct quadrature formulae with weights $\omega_{j,m}^{(i)}$, $j = 1,\ldots,m$; $i = 1,2$ such that

$$\Delta s \sum_{j=1}^{m} k(t,s)\omega_{j,m} = \int_a^b k(t,s)ds + O((\Delta s)^p)$$

and hence

$$\Delta s \sum_{j=1}^{m} k(t,s_j)\ (\omega_{j,m}^{(1)}-\omega_{j,m}^{(2)}) = O((\Delta s)^p)\ .$$

In other words, the conditioning may be very bad when the kernel is smooth.

The prospects for obtaining reasonable numerical solutions by direct quadrature do not look very good at this point. There are however schemes and equations for which this is not an unreasonable procedure. Consider for example the midpoint

$$\Delta t \sum_{j=1}^{n-1} k(t_{r+\frac{1}{2}},t_{j+\frac{1}{2}})f_j = g(t_{r+\frac{1}{2}}), \qquad r = 1,\ldots,n-1 \tag{4.3}$$

and the product midpoint schemes

$$\sum_{j=1}^{n-1} \int_{t_j}^{t_{j+1}} k(t_{r+\frac{1}{2}},s)dsf_j = g(t_{r+\frac{1}{2}})\ , \qquad r = 1,\ldots,n-1 \tag{4.4}$$

applied to the equation

$$\int_0^1 \frac{(1-\gamma^2)f(s)}{1+\gamma^2-2\gamma\cos(2\pi(t+s))}\,ds = g(t)$$

where $t_{r+\frac{1}{2}} = (r-\frac{1}{2})\Delta t$, $t_r = (r-1)\Delta t$, $\Delta t = (b-a)/(n-1)$ and $\gamma = .9$. Here, f_r are approximations to $f(t_{r+\frac{1}{2}})$, $r = 1,\ldots,n-1$. As mentioned previously, the kernel is peaked and the singular values go to zero quite slowly initially. In fact, it can be shown in this particular case that the condition number of the linear equations (4.3) and (4.4) goes to infinity like $\gamma^{-n/2}$. If we define

$$\|e_r\|_2^2 = \sum_{r=1}^{n-1} [f(t_{r+\frac{1}{2}}) - f_r]^2$$

and choose $g(t)$ so that $f(t) = \cos 2\pi t$, then we obtain the following numerical results.

	Scheme 4.3	Scheme 4.4
n	$\|e\|_2$	$\|e\|_2$
21	8.30 E-1	7.41 E-3
41	1.61 E-1	4.58 E-3
81	3.43 E-3	1.80 E-3

As expected from the peaked nature of the kernel, the product integration scheme yields superior results. In either case however a reasonable (depending of course on the application) approximation to the solution has been found. It must be stressed though that further refinement of the mesh will eventually lead to larger errors. This is due to the fact that the condition number is increasing.

A direct numerical scheme may therefore be quite adequate in some cases. For the potential problem (2.3) outlined in section two, some theoretical results are available. In de Hoog [5] stability and convergence for the product midpoint, trapezoidal and Simpson schemes is established when the contour is smooth and in Richter [14], stability and convergence is established for some Galerkin and least squares schemes. The existence of suitable direct schemes for this problem is not really very surprising since the 'conditioning' is no worse than differentiation.

However, it is not always the ill-posed nature of the problem which causes difficulties. Consider for example the problem

$$\int_0^1 (1+|t-s|^3)f(s)ds = g(t) \tag{4.5}$$

which has the solution $f(t) = g^{(iv)}(t)/12$. Let us apply the product midpoint scheme (4.4) to this equation. From analogy with finite difference approximations to fourth derivatives, we expect that the condition number of these equations will increase like n^4. Numerical results indicate that this is in fact the case but even if this could be established the convergence of the scheme (if indeed it does converge) would need further analysis as the quadrature error decreases only like n^{-2}. Nevertheless, let us apply the scheme with $n = 41$ to (4.5) where $g(t)$ is chosen so that $f(t) = e^t$. The result is indistinguishable graphically from the exact solution and is clearly quite satisfactory. Suppose now that only the $g(t_r)$, $r = 1,\ldots,n$ are given. Then, it is tempting to replace $g(t_{r+\frac{1}{2}})$ in (4.4) by the average $[g(t_{r+1}) + g(t_r)]/2$. If we implement this modified scheme we obtain the result in Figure 1 which obviously gives a poor approximation to the solution near the end points.

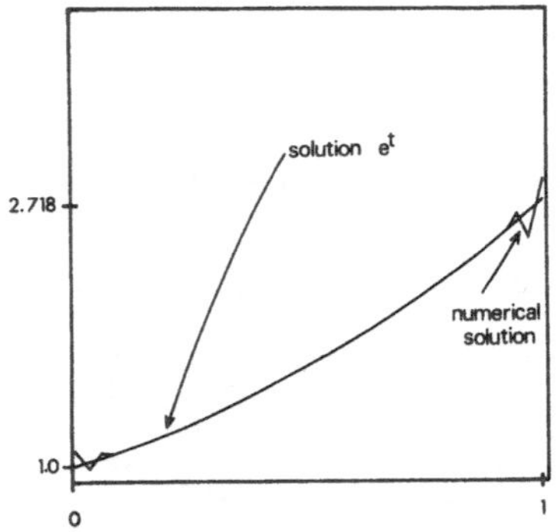

FIGURE 1

A possible explanation for this is as follows. Let us assume that the truncation error

$$\tau_r = \sum_{j=1}^{n-1} \int_{t_j}^{t_{j+1}} k(t_{r+\frac{1}{2}},s)ds f(t_{j+\frac{1}{2}}) - [g(t_{r+1})+g(t_r)]/2 , \qquad r = 1,\ldots,n-1$$

has an asymptotic expansion

$$\tau_r = (\Delta t)^2 \beta_1(t_{r+\frac{1}{2}}) + (\Delta t)^4 \beta_2(t_{r+\frac{1}{2}}) + 0\,((\Delta t)^6) .$$

Formally, we may write

$$e_r = f(t_{r+\frac{1}{2}}) - f_r$$

$$= (\Delta t)^2 \varepsilon_{1,r} + (\Delta t)^4 \varepsilon_{2,r} + 0((\Delta t)^6)$$

where $\varepsilon_{1,r}$ and $\varepsilon_{2,r}$ are the approximations obtained by applying (4.4) to

$$\int_0^1 k(t,s)\varepsilon_q(s)ds = \beta_q(t) , \qquad q = 1,2 .$$

However, in order that this equation have a solution, it is necessary that β_q , $q = 1,2$ belongs to the range of K . An explanation for the descrepancy in the numerical results could be that the lower order terms in the asymptotic expansion of the truncation error did not belong to the range of K for the modified scheme.

In the present case, a full analysis of the problem could probably be worked out as the kernel is rather simple. As the problem was constructed to illustrate a point and is not of any practical significance (at least as far as the author is aware), such an analysis is not worthwhile. However, it is clear that the verification of the extra compatability conditions will be quite difficult in general. It may well be the case that for some problems a higher order quadrature scheme (such as product integration based on piecewise quartic approximation in the present case) would have to be used for equations for which the conditioning is similar to differentiation.

Another class of direct methods can be obtained by least squares. If we use the approximation

$$f(t) \approx \sum_{r=1}^{n} f_r \eta_r(t) \equiv f_n(t) , \tag{4.6}$$

then the usual least squares procedure yields

$$\sum_{r=1}^{n} \int_a^b \beta_j(t)\beta_k(t)dt f_r = \int_a^b \beta_j(t)g(t)dt , \qquad j = 1,\ldots,n \tag{4.7}$$

where

$$\beta_j(t) = \int_a^b k(t,s)\eta_j(s)ds \ .$$

Note that when $\eta_r = \psi_r$, the eigenfunctions of K^*K , then the least squares solution corresponds to the truncated singular value solution (3.2). Such truncated solutions have been proposed by Baker *et al.* [4] and implemented by Lewis [10] who also discusses the problem of choosing a suitable value of n when the singular values go to zero very rapidly. Other basis functions such as splines can of course be used and Richter [14] has established a number of interesting stability and convergence results using splines for problems that are as badly posed as differentiation. Two disadvantages of the least squares approach are that the conditioning of the linear equation can be expected to be much worse than for the direct schemes considered previously and that the integrals on the right hand side of (4.7) cannot be calculated if the data is given at a discrete set of points. On the other hand, the consistency difficulties mentioned above are no longer a problem although they may reappear if quadrature is used to evaluate the abovementioned integrals. If a Galerkin method is used, the conditioning difficulties of the least squares approach can be avoided to some extent. Here, the approximation (4.6) is still used but the linear equations defining f_r , $r = 1,\dots,n$ are obtained by requiring that the residual is orthogonal to η_r , $r = 1,\dots, n$.

So far we have considered only schemes for which g is given analytically or is given with a high degree of accuracy on a grid of points. If f is given on a grid of points as data which contains a substantial amount of noise, we cannot expect direct schemes to work satisfactorily even if the singular values go to zero quite slowly. In the next section, we consider schemes that attempt to overcome this difficulty.

5. Stabilization of First Kind Equations

It is clear from the previous section that direct schemes do not always yield satisfactory results even if the data is given analytically or with a high degree of accuracy on a grid. If g is given as data which contains a substantial amount of noise, then (1.1) has to be stabilized in some way. Some techniques for doing this are given below.

a) Quadratic and Linear Programming Appraoch

If, from physical grounds, additional information is known about the solution, it makes sense to incorporate it into the numerical solution. Such information could include :

(i) the end point passes through a known value (for example, $f(0) = 0$) ;

(ii) the solution is bounded above or below (for example, $f(t) \geq 0$) ;

(iii) the solution is monotonically increasing or decreasing (for example $f'(x) \geq 0$) ;

(iv) the solution is concave or convex (for example, $f''(x) < 0$) .

In order to incorporate these conditions, we could attempt to minimize $\|Kf-g\|$ subject to (i) - (iv). If the problem is descretized we obtain the numerical scheme

$$\text{minimize } \|\Delta s K W \underset{\sim}{f} - \underset{\sim}{g}\|$$

subject to

$$B_1 \underset{\sim}{f} = \underset{\sim}{d}_1 \quad , \qquad B_2 \underset{\sim}{f} \geq \underset{\sim}{d}_2$$

which is a quadratic programming problem. This approach has been used quite successfully by Aurela and Torsti [2] and Torsti [19] for solving the crystal lattice vibration spectrum and the cosmic ray muon spectrum.

Another approach based on linear programming has recently been given by Babolian and Delves [3]. Here, the approximation

$$f(t) \simeq \sum_{r=1}^{n} f_r \psi_r(t)$$

is used and the quantities constrained are the growth of the coefficients f_r.

b) Regularization

By far the most popular method of stabilizing Fredholm integral equations of the first kind is the method of regularization. Simple versions of this technique were first proposed independently by Phillips [15] and Tikhonov [18].

Although there are a number of ways of justifying the mathematical formulation of this technique, we shall consider only the following. Due to the presence of noise and the possible inadequacy of our mathematic model we don't attempt to satisfy (1.1) exactly. Instead, we admit as possible solutions all functions f which make the residual small in some predetermined sense. For example we could have

$$\|Kf-g\| \leq \sigma \;. \tag{5.1}$$

Clearly, additional constraints are required so that a solution may be uniquely determined. We now seek the f satisfying (5.1) which minimizes $\|Ly\|$ where L is a linear operator. Often, $Lf = f$, $\dot{f}$ or $\ddot{f}$ is used and reflects the fact that a 'smooth' numerical solution is sought. On introducing the Lagrangian multiplier $1/\alpha$, we obtain the formulation

$$\min_{f \in D} \{\|Kf-g\|^2 + \alpha\|Lf\|^2\} \;. \tag{5.2}$$

It is interesting to note that (5.2) could have been derived by considering

$$\min_{f \in D} \|Kf-g\|$$

subject to the constraint

$$\|Ly\| \le b$$

which is more in the spirit of the quadratic programming approach outlined previously.

To illustrate a particular case of regularization, let $Lf = f'$ and $\|\cdot\| = \|\cdot\|_2$ Then, using calculus of variation, we are led to the integro-differential equation

$$- \alpha f'' + K^*Kf = K^*g$$

(5.3)

$$f'(a) = f'(b) = 0 \ .$$

The regularization parameter α is usually very small and, in view of the results for direct schemes where different discretizations of the equations could yield very different results, we might expect that the way in which (5.3) was discretized may be crucial. This does not appear to be the case and no significant differences could be detected between the Simpson and midpoint schemes when applied to a variety of problems.

There are however a number of other features about the numerical solution of (5.3) which are of interest. To illustrate these, consider the problem

$$\int_0^1 e^{ts} f(s)ds = g(t) \ , \qquad 0 \le t \le 1$$

where g has been chosen so that $f(t) = 3 \cos 3t - \cos t$. The data used was $g_r = g(r/40) + \cdot 1 \times \varepsilon_r$, $r = 0,\ldots,40$ where the ε_r are pseudo normally distributed random variables with mean zero and variance 1. Furthermore, the numerical experiment was repeated twenty times with a different random noise for each replication. One of the striking features of the numerical solutions was the variation in the value of the optimal regularization parameter α . This is illustrated in Figure 2 where the error for two replicates has been plotted against α . In both cases very good results are achieved when α is optimal but unfortunately the optimal α's are not very close. This suggests that a strategy based on choosing the regularization parameter a-priori may not be very satisfactory in general.

Another feature of the results was the variation in the solution. In Figures 3 and 4 the 'optimal' (with respect to α) solution is plotted for replicates 7 and 14 respectively.

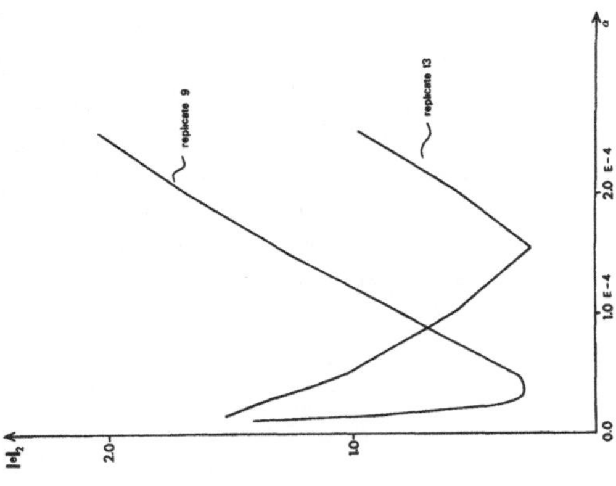

FIGURE 2

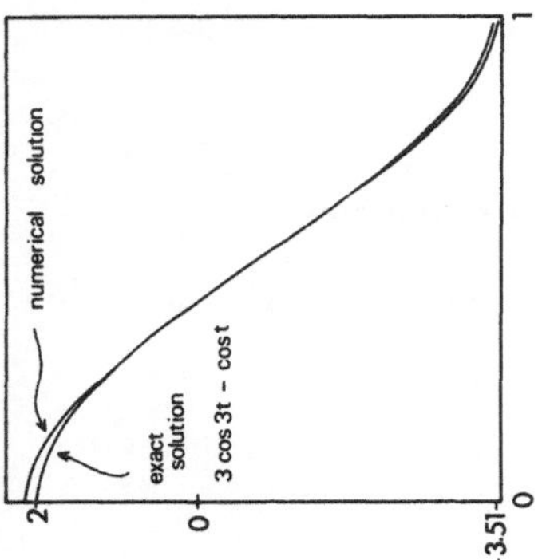

FIGURE 3

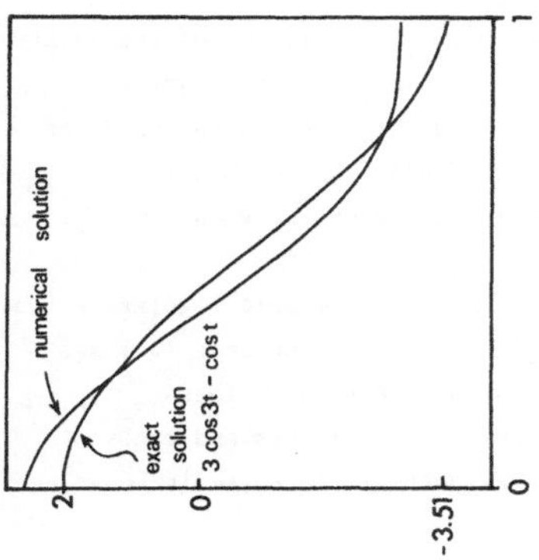

FIGURE 4

Figure 3 is quite typical and represents a satisfactory solution. However, out of the twenty replicates, about three had a similar error to that in Figure 4. Thus, there is quite a large variation in the numerical solution even when the noise is distributed in the same way. In most of the literature, one trial is usually deemed to be sufficient when comparing numerical schemes. From our limited experience, such comparisons are not very convincing.

The main difficulty in implementing regularization is choosing a suitable regularization parameter α if a suitable solution cannot be chosen visually (as may be the case in image enhancement for example). This problem is discussed by M. Lukas elsewhere in these proceedings. Among the strategies available are maximum likelihood estimates and cross validation (see for example Wahba [21]). Of course, if the variance of the noise is known, then we can solve

$$\min_{f \in D} \|Lf\|$$

subject to

$$\|Kf-g\|_2 = \sigma$$

directly (cf. Jennings [9]), but such information is not usually available.

Another approach is to generate a sequence of approximations using a variant of the Landweber iteration. For $Lf = f'$, this iteration is

$$K^*Kf_{n+1} - \alpha f''_{n+1} = K^*g - \alpha f''_n$$

$$f'(a) = f'(b) = 0 .$$

However the iteration will become unstable if applied too often. A decision when to terminate it must therefore be made.

c) Probabilistic Methods

In this framework, f and g are viewed as stationary random processes and the problem is that of finding certain statistical information about f (such as its mean on a grid of points and covariance matrix) when this information is given about the data g . To solve this problem, *a-priori* information about f is required. Such *a-priori* knowledge is really a stabilization of the problem. In fact, the schemes obtained are very similar to the zero order regularizations (i.e. $Lf = f$) considered previously. For further details see Turchin et al [20].

6. Concluding Remarks

It is clear from the preceding sections that the numerical solution of Fredholm integral equations of the first kind is a difficult task. To a certain extent, we have concentrated on the pitfalls and glossed over the fact that in practice, many people have solved such equations to their satisfaction. Nevertheless, all the methods outlined can give difficulties and extreme care must be used when using them.

Acknowledgement : The author gratefully acknowledges the assistance of Andrew Chin with the preparation of the figures.

Reference

[1] Andrews, H.C., Tescher, A.H. and Kruger, R.P., Image processing by digital computer, *IEEE Spectrum*, 9 (July 1972) pp 20-32.

[2] Aurela A.M. -nd Torsti, J.J., Correction for scattering in cosmic-ray muon spectrometers, *Ann. Acad. Sci. Fenn., Ser A*, 7 1967.

[3] Babolian, E. and Delves, L.M., An augmented Galerkin Method for first kind Fredholm equations, *Res. Rept.* CSS/78/5/1, Dept. Comp. and Stat. Sci., Univ. of Liverpool, 1978.

[4] Baker, C.T.H., Fox, L., Mayers, D.F. and Knight, K. Numerical solution of Fredholm integral equations of the first kind, *Comput. J.*, 7 (1964) pp 141-147.

[5] de Hoog, F.R. Product Integration for the Numerical Solution of Integral Equations, Ph.D thesis, Aust. Nat. Univ. 1973.

[6] Hufnagel, R.E., and Stanley, N.R. Modulation transfer function associated with image transmission through turbulent media, *J. Opt. Soc. Amer.* 54, (1964) pp 52-61.

[7] Hunt, B.R., and Janney, D.H. Digital image processing at Los Alamos scientific laboratory, *Computer*, 7 (May 1974) p 57-62.

[8] Ivanov, V.K. On Ill posed Problems, *Matem. Sb.*, 61 No.2, (1963) pp 187-199.

[9] Jennings, L.S. Orthogonal Transformations and Improperly Posed Problems. Ph.D. thesis, Aust. Nat. Univ. 1973.

[10] Lewis, B.A. On the numerical solution of Fredholm integral equations of the first kind, *JIMA*, 16, (1975), pp 207-220.

[11] Liht, M.K. The solution of minimizing a quadratic functional with approximate input data, *Ž. Uyčisl. Mat. i Mat. Fiz.* 9 (1969) pp 1004-1014.

[12] Liskovec, O.A. Regularization, of ill posed problems and a connection with the method of quasi solutions; *Differencial'nye Uravnenija* 5 (1969) 1836-1847.

[13] Mueller, P.F. and Reynolds, G.O. Image restoration by removal of random media degradations, *J. Opt. Soc. Amer.*, 57 (1967) pp 1338-1344.

[14] Richter, G.R. Numerical solution of integral equations of the first kind with non-smooth kernels, *SIAM. J. Numer. Anal.*, 15 (1978), pp 511-522.

[15] Phillips, D.L. A technique for the numerical solution of certain integral equations of the first kind, *J. Assoc. Comp. Mach.* 9, 1962, pp 84-96.

[16] Smithies, F., Integral Equations, Cambridge Univ. Press, London, 1958.

[17] Smithies, F. The eigenvalues and singular values of integral equations, *Proc. London Math. Soc.*, 43 (1937), pp 255-279.

[18] Tikhonov, A.N. On the solution of improperly posed problems and the method of regularization, *Soviet Math.* 5 (1963) pp 1035-1038.

[19] Torsti, J.J. Inversion of heat capacity by quadratic programming in the cases of K , Cℓ and Cu , *Ann. Acad. Sci. Fenn.*, *Ser A*, 2, 1972.

[20] Turchin, V.F., Kozlov, V.P., and Mlakevich, M.S. The use of mathematical statistics methods in the solution of incorrectly posed problems, *Soviet Phys. Usp.* 13, pp 681-702.

[21] Wahba, G., Practical approximate solutions to linear operator equations when the data is noisy, *Statistics Dept. Tech. Rep.* #430, Univ. of Wisconsin - Madison, (1976).

[22] Weyl, H. Das asymptotische verteilungsgestz der eigenwerte lineare partieller differentialgleichungen, *Math. Ananlen.*, 71 (1912), pp 441-479.

THE DIRECT BOUNDARY INTEGRAL METHOD FOR POTENTIAL THEORY AND ELASTICITY

Leigh J. Wardle
Division of Applied Geomechanics, CSIRO, Melbourne, Victoria

1. Introduction

This paper introduces the mathematical background to the direct boundary integral equation method, i.e., a method by which the problem of the solution of some governing equation (usually a partial differential equation) valid in a given domain is recast into the solution of an integral equation which applies only to the boundary of the domain and incorporates the boundary conditions directly. The principal advantage of such a reformulation is that the dimensionality of the problem is reduced by one. For example, a three-dimensional partial differential equation is replaced by a two-dimensional integral equation.

For many problems, the method possesses a similar level of generality to the finite element method, e.g., no limitations on the shape of the boundary or on the connectivity of the domain it encloses.

The formulation of the integral equations is demonstrated for two fundamental partial differential equations; Laplace's equation, and the equation of elasticity. Laplace's equation arises in connection with gravitational fields, electrostatic fields, steady-state heat conduction, incompressible flow, torsion, etc.

Fredholm integral equations follow from the representation of harmonic functions by single-layer or double-layer potentials. The existence of solutions to such equations was first demonstrated by Fredholm (1903) using a discretization procedure. However, the use of discretization processes to actually construct solutions did not become feasible until the advent of electronic digital computers.

An attractive alternative to the "classical" source density approach (i.e., the use of single-layer or double-layer potentials) is the main basis of this paper. For the potential theory example, this involves exploiting Green's formula, which represents a harmonic function as the sum of single-layer and double-layer potentials. By taking the field point on the boundary, a boundary integral equation is obtained that only involves the actual physical boundary data, i.e., values of the harmonic function and its normal derivative. For this reason we call the method the *direct* boundary integral equation method, thus distinguishing it from the source density method which is *indirect* because the unknown boundary data can only be obtained after the source density has been solved for.

The formulation of the fundamental boundary value problems of linear elasticity by vector integral equations analogous to the Fredholm integral equations of potential theory was introduced by Kupradze (1965). Kupradze's formulation is an indirect one, i.e., the unknowns in the integral equations are vector source densities. The formulation of direct boundary integral equations was introduced by Rizzo (1967) for plane problems and Cruse (1969) for three-dimensional problems.

This paper illustrates the essential features of the direct boundary integral equation method by formulating the integral equations for potential theory and elasticity. An overview of the numerical procedures used to solve the integral equations is given. Brief details are also given of methods used to handle non-homogeneous and non-elliptic partial differential equations and problems involving non-homogeneous material properties.

2. Potential Theory

The formulation of the direct boundary integral equation for the solution of Laplace's equation is based on the existence of two items:

(a) Fundamental Solution to Laplace's Equation: A singular solution of Laplace's equation with the Dirac delta function for a discrete inhomogeneity.

(b) Green's Theorem: A reciprocal relationship between two functions which have continuous first derivatives.

2.1 FUNDAMENTAL SOLUTIONS

The Laplacian operator in three dimensions is $\partial^2/\partial r^2 + (2/r)\,\partial/\partial r$ and in two dimensions is $\partial^2/\partial r^2 + (1/r)\,\partial/\partial r$. It is easily verified by direct substitution that the functions $1/r(p,q)$ and $\log[1/r(p,q)]$ satisfy the respective Laplace's equation for $r(p,q) \neq 0$, where $r(p,q)$ is the scalar distance between points p and q.

Furthermore, if the point $r(p,q) = 0$ is included it can be shown that for three dimensions

2.1 $$\nabla^2 \frac{1}{r(p,q)} = -4\pi\, \delta(p,q)$$

where the Dirac delta function $\delta(p,q)$ is defined to have the following properties

2.2 (i) $$\delta(p,q) = 0 \ , \quad \underset{\sim}{x}_p \neq \underset{\sim}{x}_q$$

2.3 (ii) $$\int_R \rho(q)\ \delta(p,q)\ dV(q) = \rho(p)\ .$$

We infer the function

2.4 $$\psi(p,q) = \frac{1}{r(p,q)}$$

to be the fundamental solution in three dimensions. The fundamental solution in two dimensions is

2.5 $$\psi(p,q) = \log \frac{1}{r(p,q)}\ .$$

2.2 GREEN'S THEOREM

If ϕ and ψ are two continuous functions with continuous first derivatives, the application of the divergence theorem to the region R with surface S gives the following Green's theorem:

2.6 $$\int_R (\phi\nabla^2\psi - \psi\nabla^2\phi)dV = \int_S (\phi\,\frac{d\psi}{dn} - \psi\,\frac{d\phi}{dn})dS\ .$$

The normal direction to S is taken outwards from R.

We take $\phi(q)$ to be our *unknown* harmonic function (satisfies $\nabla^2\phi = 0$) and $\psi(p,q)$ as the *fundamental* solution to Laplaces equation (from Section 2.1).

As ψ does not obey Laplace's equation at $p(x)$, we surround this point by a small sphere (or circle) of radius ε, denoted R_ε with surface S_ε. In the region $R-R_\varepsilon$ $\nabla^2\psi = 0$. Taking Green's theorem for this region

2.7 $$\int_{R-R_\varepsilon} (\phi\nabla^2\psi - \psi\nabla^2\phi)dV = \int_{S+S_\varepsilon} (\phi\,\frac{d\psi}{dn} - \psi\,\frac{d\phi}{dn})dS\ .$$

On the surface S_ε $d/dn = -d/dr$; replacing $\psi(p,q)$ by the appropriate fundamental solution, the integral on the surface S_ε becomes

2.8 $$\lim_{\varepsilon\to 0} \int_{S_\varepsilon} \phi\,\frac{d\psi}{dn} = 2\alpha\pi\phi(p)$$

where $\alpha = 2$ for three-dimensional problems and $\alpha = 1$ for two-dimensional problems. In the limit as $\varepsilon \to 0$ equation (2.7) becomes

2.9 $$\phi(p) = \frac{1}{2\alpha\pi}\int_S -\phi(Q)\,\frac{d\psi(p,Q)}{dn(Q)} + \frac{d\phi(Q)}{dn(Q)}\,\psi(p,Q)\,ds(Q)\ .$$

This is the important integral identity of potential theory, that a harmonic function ϕ may be expressed as the sum of a single-layer potential with density $(1/2\alpha\pi)\ d\phi/dn$ and a double-layer potential with density $(-1/2\alpha\pi)\phi$.

2.3 DIRECT BOUNDARY INTEGRAL EQUATION

We now consider equation (2.9) as the interior point $P(\underset{\sim}{x})$ is taken to a boundary point $P(\underset{\sim}{x})$.

The function $\phi(p)$ is assumed continuous so that $\phi(p) \to \phi(P)$. The second term in equation (2.9) is also continuous if $d\phi/dn$ is bounded. The first term, however, represents a double-layer potential with density $\phi(Q)$ and has the jump property given by

2.10 $$\phi^{+}(P) = -\alpha\pi\phi(P) + \int_S \phi(Q)\,\frac{d\psi(P,Q)}{dn(Q)}\,dS(Q)$$

where $\phi^{+}(P)$ represents the limit as $p(\underset{\sim}{x}) \to P(\underset{\sim}{x})$ from inside R. The integral in equation (2.10) must be interpreted in the Cauchy Principal Value sense.

Thus, in the limit, equation (2.9) becomes

2.11 $$\alpha\pi\phi(P) + \int_S \phi(Q)\,\frac{d\psi(P,Q)}{dn(Q)}\,dS(Q) = \int_S \psi(P,Q)\,\frac{d\phi(Q)}{dn(Q)}\,dS(Q)\ .$$

Equation (2.11) represents a constraint equation between the Dirichlet boundary conditions (ϕ defined) and the Neumann ($d\phi/dn$ defined) boundary conditions. For Neumann boundary conditions the right hand side of equation (2.11) is known, giving a Fredholm equation of the second kind for the unknown boundary values of the function $\phi(Q)$. For the Dirichlet problem, equation (2.11) becomes a Fredholm equation of the first kind for the unknown boundary values of $d\phi/dn$. The mixed boundary value problem leads to a mixed integral equation for the unknown boundary data.

As the unknowns in the integral equation are physical boundary quantities (either ϕ or $d\phi/dn$), equation (2.11) is called the *direct* boundary integral equation to distinguish it from integral equations that involve a 'fictitious' source density. Formulations based on the source density approach are called *indirect* because the 'complementary' boundary data can only be obtained after the source density distribution has been solved for.

2.4 REGULARITY OF INTEGRAL OPERATORS

The kernel $K(P,Q)$ of an integral operator is the term multiplying the density under the integral sign. A kernel is defined as being singular if

2.12 $$\lim_{Q \to P} r^{\alpha}(P,Q)K(P,Q) \neq 0$$

where $\alpha = 1$ for two-dimensional problems and
$\alpha = 2$ for three-dimensional problems.

For the most singular kernel introduced in the discussion of potential theory $d\psi/dn$, the left side of (2.12) reduces to

$$\lim_{Q \to P} \frac{dr(P,Q)}{dn(Q)}$$

which approaches zero uniformly in the limit. Thus, the kernels ψ, $d\psi/dn$ are *non-singular*.

Thus, for the Neumann boundary conditions, the Fredholm alternative theorem can be used to prove the existence and uniqueness (to within a constant) of a solution. For the Dirichlet boundary conditions, the fact that a Fredholm equation of the first kind is involved (i.e., Fredholm alternative theorem does not apply) is not a drawback because for a well-posed problem the existence of a solution can be expected on physical grounds. Moreover, the solution can be shown to be unique.

3. Elasticity Theory

3.1 INTRODUCTION

This section describes the formulation of a *direct* boundary integral equation method for linear static isotropic elasticity.

The development for elasticity is described to show that not all the features encountered in the potential theory apply. In particular, unlike potential theory the integral equations will sometimes be *singular*, but will be subject to the Fredholm alternative theorems. Moreover, the boundary integral equations are coupled sets of vector equations as contrasted with the scalar equations of potential theory. Finally, while the displacement boundary value problem is a Dirichlet problem, the more usual traction boundary value problem is analogous to, but is not, a Neumann problem.

3.2 GOVERNING EQUATIONS

The governing equations are

(i) *the stress equilibrium equations*

3.1 $$\sigma_{ij,j} + X_i = 0$$

where σ_{ij} is the stress tensor and X_i is the body force vector.

(ii) *the strain-displacement relations*

3.2 $$\varepsilon_{ij} = \tfrac{1}{2}(u_{j,j} + u_{j,i})$$

where ε_{ij} is the strain tensor.

(iii) *the stress-strain relations ("Hooke's Law")*

3.3 $$\sigma_{ij} = \frac{2\mu\nu}{1-2\nu}\,\delta_{ij}\,\varepsilon_{mm} + 2\mu\,\varepsilon_{ij}$$

where ν is Poisson's ratio of material and μ is shear modulus of material.

Substitution of equations (3.2) and (3.3) into equation (3.1) results in Navier's equations of equilibrium in terms of the displacements

3.4 $$\frac{1}{1-2\nu}\,u_{i,ij} + u_{j,ii} + \frac{1}{\mu}\,X_j = 0\ .$$

These equations form an elliptic system (Birkhoff, 1971).

For most boundary value problems of interest, the body force term X_j is zero. Equation (3.4) must be solved subject to certain boundary conditions. The displacement boundary value problem assumes knowledge of the displacements on the entire surface S

3.5 $$u_i(\underset{\sim}{x}) = g_i(\underset{\sim}{x})$$

The traction (i.e., stress resultant) boundary value problem is more complicated. The traction vector is defined by

3.6 $$t_i = \sigma_{ij}\,n_j = h_i(\underset{\sim}{x})\ .$$

The vector function $h_i(\underset{\sim}{x})$ is known for $\underset{\sim}{x}$ in S and n_j is the outwards normal to S at $\underset{\sim}{x}$.

Substituting equations (3.2) and (3.3) gives

3.7 $$\frac{du_i}{dn} + u_{j,i}\,n_j + \frac{2\nu}{1-2\nu}\,u_{j,j}\,n_i = \frac{1}{\mu}\,h_i(\underset{\sim}{x})\ .$$

Equation (3.5) is a Dirichlet-type boundary condition. However, although the first term in equation (3.7) is a Neumann-type boundary condition, the other terms eliminate the direct relation between the Neumann boundary conditions of potential theory and traction boundary conditions of elasticity.

3.3 FUNDAMENTAL SOLUTIONS

The fundamental solution u_i^* for the Navier equations (3.4) is the well-known Kelvin solution (Sokolnikoff, 1956) for a point load (i.e., the body-force term X_i is a Dirac delta function) in an infinite body.

This solution can be written in the form

3.8 $$u_i^*(p,q) = U_{ji}(p,q)e_j$$

and the tractions on an arbitrary surface around the point $p(\underset{\sim}{x})$ are given by

3.9 $$t_i^*(p,q) = T_{ji}(p,q)e_j \ .$$

The e_j's are a set of unit vectors in the x_j directions.

The tensors U_{ji} and T_{ji} are given by

3.10 $$U_{ji} = \frac{1}{r}\,[(3-4\nu)\ \delta_{ij} + r_{,i}\ r_{,j}]/16\pi\mu(1-\nu)$$

3.11 $$T_{ji} = -\frac{1}{r^2}\{\frac{dr}{dn}\,[(1-2\nu)\ \delta_{ij} + 3r_{,i}\ r_{,j}] - (1-2\nu)(n_j\ r_{,i} - n_i\ r_{,j})\}/8\pi(1-\nu)$$

for three dimensions, and

3.12 $$U_{ji} = [(3-4\nu)\ \delta_{ij}\ \log\frac{1}{r} + r_{,i}\ r_{,j}]/8\pi\mu(1-\nu)$$

3.13 $$T_{ji} = -\frac{1}{r}\,\{\frac{dr}{dn}\,[(1-2\nu)\ \delta_{ij} + 2r_{,i}\ r_{,j}] - (1-2\nu)(n_j\ r_{,i} - n_i\ r_{,j})\}/4\pi(1-\nu)$$

in two dimensions.

The normal is taken at $q(\underset{\sim}{x})$.

3.4 BETTI'S ELASTIC RECIPROCAL THEOREM

Betti's second theorem relates two independent equilibrium states u_i, t_i and u_i^*, t_i^* with zero body force (Sokolnikoff, 1956)

3.14 $$\int_S t_i u_i^*\, dS = \int_S t_i^* u_i\, dS \ .$$

To use this theorem to obtain a direct boundary integral equation we take u_i and t_i as the solution of our problem of interest and u_i^* and t_i^* the fundamental solution given by equations (3.8) and (3.9). Since the fundamental solution represents a point body force at $p(\underset{\sim}{x})$, we delete from R a sphere (or circle) of radius ε denoted R_ε surrounding $p(\underset{\sim}{x})$ with surface S_ε and rewrite equation (3.14).

3.15 $$\int_{S+S_\varepsilon} t_i u_i^* \, dS = \int_{S+S_\varepsilon} t_i^* u_i \, dS$$

It can be easily verified that

3.16 $$\lim_{\varepsilon \to 0} \int_{S_\varepsilon} U_{ji}(p,Q)\, dS(Q) = 0$$

3.17 $$\lim_{\varepsilon \to 0} \int_{S_\varepsilon} T_{ji}(p,Q)\, dS(Q) = \delta_{ij} .$$

Substituting equations (3.8), (3.9), (3.16) and (3.17) into equation (3.14) and taking each of the e_j terms as independent gives

3.18 $$u_j(p) = - \int_S u_i(Q) T_{ji}(p,Q)\, ds(Q) + \int_S t_i(Q) U_{ji}(p,Q)\, dS(Q) .$$

Equation (3.18) is known as Somigliana's identity and states that the displacement at any interior point $p(\underset{\sim}{x})$ can be expressed in terms of a single-layer vector potential of density $t_i(Q)$ and a double-layer vector potential of $-u_i(Q)$. Equation (3.18) is the elastic counterpart of the potential theory equation (2.9).

3.5 DIRECT BOUNDARY INTEGRAL EQUATION

Again, following the procedure used in Section 2.3 for potential theory, we determine the limiting form of equation (3.18) as $p(\underset{\sim}{x}) \to P(\underset{\sim}{x})$. The single-layer potential arising from the U_{ji} term is continuous as p approaches the boundary, but the double-layer potential has the jump behaviour given by

3.19 $$u_i^+(P) = -u_i(P)/2 + \int_S u_j(Q) T_{ji}(P,Q)\, dS(Q)$$

where u_i^+ represents the limit from inside R. As before, the integral must now be interpreted in the Cauchy Principal Value sense.

Substituting equation (3.19) into equation (3.18) gives

3.20 $$u_j(P)/2 + \int_S u_i(Q) T_{ji}(P,Q)\, dS(Q) = \int_S t_i(Q) U_{ji}(P,Q)\, dS(Q) .$$

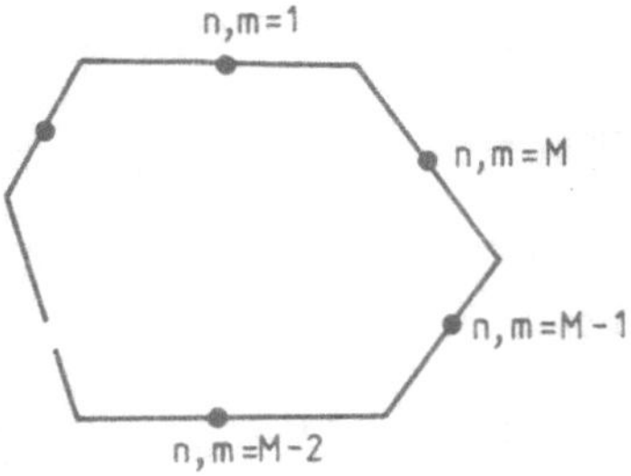

Figure 1. Simple two Dimensional Boundary Discretization Scheme

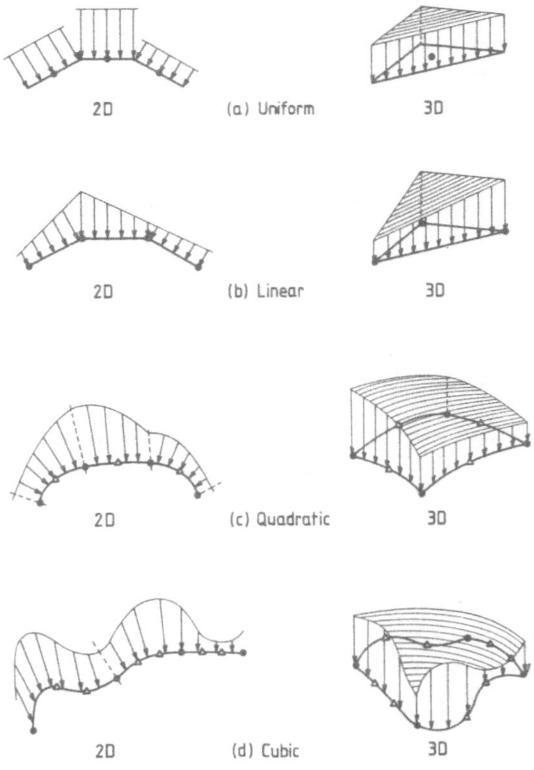

Figure 2.

The direct boundary integral equation (3.20) is a constraint equation between surface displacements and surface tractions for an elastic body.

3.6 REGULARITY OF INTEGRAL OPERATORS

The kernels $U_{ji}(Q,P)$ (equations (3.10) and (3.12)) are non-singular. The kernels $T_{ji}(Q,P)$ (equations (3.11) and (3.13) contain the term

$$T_{ij}^{*}(P,Q) = \frac{1}{r^{\alpha}}(1-2\nu)\,(n_j\, r_{,i} - n_i\, r_{,j})/4\alpha\pi(1-\nu)$$

which from equation (2.12) are *singular* kernels. However, Cruse (1977) uses Mikhlin's (1965) results to show that the kernels can be regularized and the Fredholm alternative theorems can be applied.

Since the *traction*-defined boundary value problem reduces the boundary integral equation (5.1) to a Fredholm equation of the *second* kind, involving the *singular* kernel T_{ji}, the Fredholm alternative theorem can be used to show that the non-rigid body boundary displacements are unique. For *displacement*-defined boundary conditions, equation (5.1) reduces to a Fredholm equation of the *first* kind with the *non-singular* kernel U_{ji}. The existence of a solution to this equation can be expected on physical grounds and can be shown to be unique (Cruse, 1977).

4. NUMERICAL SOLUTION OF BOUNDARY INTEGRAL EQUATION

4.1 INTRODUCTION

Except for very simple geometries, analytical solutions to the boundary integral equations are unavailable and numerical methods must be used. We seek reduction of the integral equation to an algebraic form tractable for numerical solution.

For purposes of illustration, consider the two-dimensional potential theory integral equation (equation (2.11) with $\psi = \log(\frac{1}{r})$). The simplest method of discretization is to divide the boundary of the body into M straight line segments (see Figure 1). The function and its normal derivative are assumed constant on each segment. The boundary integral equation is written at the mid-point of each segment and the values of ϕ and $d\phi/dn$ can be extracted from the integration over each segment

4.1
$$\pi\phi(P_n) + \sum_{m=1}^{M} \phi(Q_m) \int_{\Delta S_m} \frac{d}{dn(Q)} \log \frac{1}{r(P_n,Q)}\, dS(Q) =$$
$$\sum_{m=1}^{M} \frac{d\phi}{dn}\Big|_{Q_m} \int_{\Delta S_m} \log \frac{1}{r(P_n,Q)}\, dS(Q) \quad (n = 1,M)\ .$$

The integrals can be expressed in closed form (Cruse, 1977). Care must be taken to handle the segment where at the mid-point $P_n = Q_m$. This integral must be evaluated in the Cauchy Principal Value sense, i.e., by deleting a small segment of length 2ε at the mid-point and taking the limit

4.2 $$\lim_{\varepsilon\to 0} \left\{ \int_{Q_1}^{-\varepsilon} [\quad] dS(Q) + \int_{\varepsilon}^{Q_2} [\quad] dS(Q) \right\} .$$

Denoting the results of the integrals by $\Delta F^*(P_n, Q_m)$ and $\Delta G(P_n, Q_m)$, equation (4.1) becomes

4.3 $$\pi\phi(P_n) + \sum_{m=1}^{M} \phi(Q_m)\Delta F^*(P_n, Q_m) = \sum_{m=1}^{M} \frac{d\phi}{dn}\Big|_{Q_m} \Delta G(P_n, Q_m) \quad (n = 1,M) .$$

Equation (4.3) may be rewritten in matrix form (ΔF^* becomes ΔF by adding π to the diagonal terms of ΔF)

4.4 $$[\Delta F]\ \{\phi\} = [\Delta G]\ \{\frac{d\phi}{dn}\} .$$

For any well-posed boundary value problem, equation (4.4) involves just M knowns and M unknowns. This equation is normally very well conditioned and can be solved by standard matrix reduction schemes.

Values of ϕ at any interior point can be obtained using a discretized version of equation (2.9).

4.2 HIGHER-ORDER SHAPE FUNCTIONS

The crude discretization approach described in Section 4.1 has been supplanted by the use of more sophisticated 'shape-functions' for the functional variation along each boundary segment. The next step up in complexity is to use linear variation on each segment (Figure 2b). For this case, the unknowns in the system are values at the 'nodes', i.e., the intersections of the straight line segments. The three-dimensional analogue involves triangular elements with linear variation between the vertices.

More recent implementations have seen the introduction of quadratic (Figure 2c) and cubic (Figure 2d) variation of boundary data. To enable efficient modelling of curved boundaries, the same functional variation has been used for the boundary surface (i.e., "isoparametric" elements). These elements have been used by Wu et al. (1977) for potential theory and Lachat and Watson (1976) for elasticity. A somewhat theoretical analysis of the use of curved elements is given by Nedelec (1976).

5. Further Applications

5.1 POISSON'S EQUATION

Consider Poisson's equation with an assumed continuous right-hand side

5.1 $$\nabla^2\phi(\underset{\sim}{x}) = \rho(\underset{\sim}{x}) .$$

If the term $\rho(x)$ is included in Green's theorem (Equation (2.6)), the boundary integral equation (2.11) becomes

5.2 $$\alpha\pi\phi(P) = \int_S \phi(Q)\,\frac{d\psi(P,Q)}{dn(Q)}\,dS(Q) = \int_S \psi(P,Q)\,\frac{d\phi(Q)}{dn(Q)}\,dS(Q) + \int_R \rho(q)\psi(P,q)\,dV(q) .$$

We now have a known volume integral term involving $\rho(q)$, and for arbitrary $\rho(q)$ we must resort to a volume discretization scheme to evaluate the volume integral. The boundary integral equation method thus loses its main advantage, i.e., only the boundary has to be discretized. However, if the right hand side of Poisson's equation is harmonic i.e., $\nabla^2\rho(x) = 0$, Wu et al. (1977) show that for the two-dimensional case the volume integral can be expressed as a simple surface integral.

5.2 NON-ZERO BODY FORCES IN ELASTICITY

When non-zero body forces X_i are involved, the governing partial differential equation (3.4) becomes non-homogeneous, and in similar fashion to the potential theory case, the boundary integral equation (3.20) must include a volume integral involving X_i and U_{ji}.

Stippes and Rizzo (1977) show that if

$$X_i = -\psi_{,i}$$

with $\nabla^2\psi = k_o$ (a constant)

then the volume integral can be reduced to a surface integral. These conditions are satisfied by a constant gravitational field, a centrifugal force arising from fixed axis rotation, or the body forces induced by a steady-state thermal field.

5.3 NON-HOMOGENEOUS MATERIAL PROPERTIES

An implicit assumption used in the formulation of the boundary integral equations is that the material properties are uniform throughout the domain. This stems from the fact that the fundamental solutions involved are for uniform material properties. For a piecewise-homogeneous system, e.g., a steady-state heat conduction problem involving distinct zones of different thermal resistivities, the boundary integral equation can be written for each zone. Discretization of the boundary integral equation for each zone leads to a system of linear equations.

These linear equations can be assembled into a system representing the behaviour of the composite body by applying suitable continuity conditions on the interfaces between the regions. Thus, the implementation of the boundary integral equation method to this class of problem reduces to one of computer programming rather than sophisticated mathematics. The application of this approach to potential theory problems is described by Butterfield and Tomlin (1972) and to elasticity problems by Wardle and Crotty (1978).

5.4 MORE GENERAL ELLIPTIC SYSTEMS

Consider the linear elliptic system

$$L_{ij}(u_j) = 0 \quad \text{in } R$$

where L_{ij} is a general elliptic operator of order m with analytic coefficients.

It can be shown (John, 1955) that a set of fundamental solutions ψ_{jk} exists such that

$$u_j(p) = \int_S M_{ik}(u_i(Q),\psi_{jk}(p,Q))ds(Q) .$$

This equation is the analogue of equations (2.9) and (3.18). As before, the direct boundary integral equation is obtained by taking the limit as $p \to P$.

5.5 NON-ELLIPTIC SYSTEMS

The applications described so far generally arise from the investigation of steady-state phenomena. When transient phenomena are considered the complexity of the system is increased by the independent variable of time.

The most commonly used technique is to temporarily eliminate time as an independent variable by Laplace transforming the partial differential equations and boundary conditions. In this manner, parabolic and hyperbolic systems generally reduce to elliptic systems. The elliptic system can be solved in the transform space by the direct boundary integral equation method for a sequence of values of the transform parameter. The time solution can then be effected by numerical transform inversion. A survey of the application of this technique to transient heat conduction, viscoelasticity and wave propogation is given by Shippy (1975).

6. CONCLUSIONS

For an increasing number of applications the direct boundary integral equation method is an important alternative to the conventional finite element method, i.e., the finite element method applied to the partial differential equations in the domain. However, as can be seen from the discussion of numerical techniques, we are in fact using a finite element method for the boundary integral equation. Perhaps it is better to speak of boundary elements and volume elements.

The direct boundary integral equation method exploits to the maximum the fundamental relations between boundary and solution functions. Rather than discretizing everything at the outset, the method goes as far as possible analytically and then introduces approximations that are straightforward and effective.

7. References

[1] Birkhoff, G. The Numerical Solution of Elliptic Equations, Society for Industrial and Applied Mathematics, Philadelphia, 1971.

[2] Butterfield, R. and Tomlin, G.R. Integral techniques for solving zoned anisotropic continuum problems, International Conference on Variational Methods in Engineering, September 25-29, Department of Civil Engineering, Southampton University, England, Session 9, 9/31-9/51, 1972.

[3] Cruse, T.A. Numerical solutions in three-dimensional elastostatics, *Int. J. Solids & Structures*, 5, 1259-1274, 1969.

[4] Cruse, T.A. Mathematical foundations of the boundary-integral equation method in solid mechanics, Air Force Office for Scientific Research TR-77-1002, 1977.

[5] Fredholm, I. Sur une classe d'equations fonctionelles, *Acta Math.*, 27, 365-390, 1903.

[6] John, F. Plane Waves and Spherical Means, 1st ed., *Interscience*, New York, 1955.

[7] Kupradze, V.D. Potential Methods in the Theory of Elasticity, (Daniel Davey and Company, New York), 1965.

[8] Lachat, J.C. and Watson, J.O. Effective numerical treatment of boundary integral equations A formulation for three-dimensional elastostatics, *Int. J. Numerical Methods in Engineering*, 10, 991-1005, 1976.

[9] Mikhlin, S.G. Multi-dimensional Singular Integrals and Integral Equations, Pergamon, London, 1965.

[10] Nedelec, J.C. Curved finite element methods for the solution of singular integral equations on surfaces in R^3. *Computer Methods Appl. Mech. Eng.*, 8, 61-80, 1976.

[11] Rizzo, F.J. An integral equation approach to boundary value problems of classical elastostatics, *Quarterly of Applied Mathematics*, 25, No.1, 83-95, 1967.

[12] Shippy, D.J. Application of the boundary integral equation method to transient phenomena in solids, Boundary Integral Equation Method, *Comp. Appl. in Applied Mechanics*, T.A. Cruse and F.J. Rizzo (eds), ASME New York, 15-30, 1975.

[13] Sokolnikoff, I.S. Mathematical Theory of Elasticity, 2nd ed., McGraw-Hill, New York. 1956.

[14] Stippes, M. and Rizzo, F.J. Note on body force integral of classical elastostatics, *Zeitschrift fur Angewandte Mathematik und Physik*, 28, 339-341, 1977.

[15] Wardle, L.J. and Crotty, J.M. Two-dimensional boundary integral equation analysis for non-homogeneous mining applications, In: *Recent Advances in Boundary Element Methods*, (ed. C.A. Brebbia), Pentech Press, London, 233-251, 1978.

[16] Wu, Y.S., Rizzo, F.J., Shippy, D.J. and Wagner, J.A. An advanced boundary integral equation method for two-dimensional electro-magnetic field problems, *Electrical Machines and Electro-mechanics*, 1, No.4, 301-313, 1977.

REGULARIZATION

M.A. Lukas
Computing Research Group/Pure Mathmatics, SGS,
Australian National University, Canberra, A.C.T.

1. Introduction

For the approximate solution of the first kind Fredholm integral equation

$$Ku(x) = \int_0^1 K(x,t)u(t)dt = f(x) \quad x \in [0,1] , \tag{1.1}$$

a major difficulty when the kernel $K(x,t)$ is smooth is the sensitivity of its solution u_0 to small perturbations in f. Because of this property, (1.1) is often referred to as improperly posed. To cope with this numerically, it is desirable to replace (1.1) by some stabilized problem, in operator form,

$$K_\alpha u_\alpha = f , \qquad \alpha > 0$$

which satisfies the following properties

(i) $u_\alpha \to u_0$ as $\alpha \to 0$

(ii) the sensitivity of u_α to small perturbations in f decreases as α increases.

One such approach which has proved to be effective for a wide class of problems, is regularization.

In regularization, the approximate solutions u_α are defined variationally by

$$\underset{u \in W}{\text{minimize}} \ \|Ku-f\|^2 + \alpha\Omega(u) , \tag{1.2}$$

where W is some space of smooth functions and $\alpha > 0$ is a constant, called the regularization parameter. Here Ω is some non-negative "stabilizing" functional which controls the sensitivity of the regularized solution u_α to perturbations in f. Examples are $\Omega(u) = \|u\|_2^2$ or $\Omega(u) = \|u''\|_2^2$ where $\|\cdot\|_2$ is the norm on $L^2 = L^2(0,1)$. The types of norm in (1.2) that will be considered are set out below.

The integral operator K is assumed to be bounded $L^2 \to L^2$. A sufficient condition for this is that the kernel $K(x,t)$ belong to $L^2((0,1) \times (0,1))$. This is a very mild restriction since regularization has most importance for integral equations with smooth kernels.

It is clear that (1.2) defines a tradeoff between minimizing the residual and stabilizing the solution, and the parameter α determines the weighting between the two. If α is too small, the original improperly posed problem (1.1) will tend to dominate, and therefore numerical errors may be unduly magnified. If, on the other hand, α is too large, the resulting solution, although quite stable, will not adequately satisfy (1.1). This suggests that there may be an optimal α. Although it is of great importance, an examination of the various criteria for the choice of α discussed in the literature will not be pursued in this paper. However, because of its prominence, the method of weighted cross-validation for choosing α will be discussed in section 7.

The major aim of this paper is to examine the dependence of the regularized solution u_α on the choice of norm $\|\cdot\|$ and stabilizing functional $\Omega(u)$. Together these determine a natural framework in which an explicit representation for u_α can be derived. It is important that the form and properties of u_α be understood before applying a particular regularization procedure. Depending on the choice of $\|\cdot\|$ and $\Omega(u)$, the following forms of regularization will be discussed in the sequel.

Tikhonov Regularization This corresponds to a choice of the L^2 norm in (1.2) along with $\Omega(u)$ as the square of a weighted Sobolev norm, namely

$$\Omega(u) = \sum_{i=0}^{m} \int_0^1 w_i(x)\left(u^{(i)}(x)\right)^2 dx\,, \quad w_i(x) > 0\,.$$

This technique will be discussed in section 2.

Regularization with Differential Operators The norm is again the L_2 norm but $\Omega(u)$ is defined by

$$\Omega(u) = \|Tu\|_2^2\,,$$

where T is a linear differential operator, e.g. $Tu = u'$. This method will be examined in section 3 and a constructive existence and uniqueness proof for u_α given.

Discrete Norm and Discrete Stabilizing Functional This may be thought of as a full discretization of both the norm and stabilizing functional with the operator K and functions

u and f now corresponding to a matrix K_n and vectors $\underset{\sim}{u}$ and $\underset{\sim}{f}$. The norm becomes a weighted Euclidean norm

$$\|K_n\underset{\sim}{u}-\underset{\sim}{f}\|^2 = \sum_{i=1}^{n} [(K_n\underset{\sim}{u})_i - f_i]^2\sigma_i$$

while the stabilizing functional becomes some discrete version of the continuous functionals above,

$$\text{e.g.} \qquad \frac{1}{n}\sum_{i=0}^{n-1}\left(\frac{u_{i+1}-u_i}{h}\right)^2 \quad \text{instead of} \quad \|u'\|_2^2 .$$

This discrete problem will be discussed in section 4.

Discrete Norm and Continuous Stabilizing Functional The norm is again a weighted Euclidean norm and the stabilizing functional $\Omega(u)$ is either a Tikhonov functional or the operator functional defined above. The problem then is to find $u_\alpha \in W$ which minimizes

$$\sum_{i=1}^{n} (Ku(x_i) - \underset{\sim}{f}_i)^2\sigma_i + \alpha\Omega(u) ,$$

where $\underset{\sim}{f}$ is some discrete approximation to f. Often in practice, f is only given discretely and so this form of regularization is important. There are two different approaches to characterizing u_α, one due to Anselone and Laurent [5] and another due to Wahba [24]. These will be examined in sections 5 and 6 and a comparison between them made.

2. Tikhonov Regularization

The technique of regularization was introduced by Tikhonov [21], [22], in the following way: for a given first kind Fredholm integral equation

$$Ku(x) = \int_0^1 K(x,t)u(t)dt = f(x) \qquad x \in [0,1]$$

replace it by the variational problem

$$\underset{u\in W}{\text{minimize}} \ \|Ku-f\|_2^2 + \alpha\Omega(u) \tag{2.1}$$

where W is some space of smooth functions and $\Omega(u)$ is the square of a weighted Sobolev norm, defined by

$$\Omega(u) = \sum_{i=0}^{m}\int_0^1 w_i(x)\left(u^{(i)}(x)\right)^2 dx , \qquad w_i \in C[0,1] \quad \text{and} \quad w_i(x) > 0 .$$

The operator $K : W \to L^2$ is assumed to be bounded. It is clear that in the minimization (2.1), $\Omega(u)$ controls the smoothness of u and thus, as will be shown, reduces the solutions sensitivity to perturbations in f.

Here it is assumed, at least initially, that f is given analytically. An application of this exists in the work of Anderssen [2] on estimating moment functionals, in which $f(x) = x^n$.

If W is taken to be the Sobolev space

$$W^{m,2}(0,1) = \{u \in L^2 : \text{weak derivatives } u^{(i)} \text{ exist for } i \le m \text{ and } u^{(m)} \in L^2\} ,$$

then the variational formulation (2.1) is equivalent to the Euler-Lagrange equation

$$K^*Ku + \alpha \sum_{i=0}^{m} (-1)^i \left(w_i u^{(i)}\right)^{(i)} = K^*f \tag{2.2}$$

with boundary conditions

$$B_j(u)|_{0,1} = \sum_{i=j}^{m} (-1)^{i-j} \left(w_i u^{(i)}\right)^{(i-j)} \Big|_{0,1} = 0 \qquad j = 1,\dots,m .$$

Here $$K^*Ku(x) = \int_0^1 \int_0^1 K(s,x)K(s,t)u(t)\,ds\,dt$$

and $$K^*f(x) = \int_0^1 K(t,x)f(t)\,dt .$$

By using the Green's function $G(x,t)$ for the boundary value problem

$$Lu = \sum_{i=0}^{m} (-1)^i \left(w_i u^{(i)}\right)^{(i)} = f$$

$$B_j(u)|_0 = B_j(u)|_1 = 0 \qquad j = 1,\dots,m ,$$

which exists if $w_i(x) > 0$, Tikhonov transforms (2.2) into the second kind Fredholm integral equation

$$(GK^*K+\alpha I)u = GK^*f , \tag{2.3}$$

where $$Gu(x) = \int_0^1 G(x,t)u(t)\,dt .$$

Since $GK^*K : L^2 \to L^2$ has no negative eigenvalues, it follows from the Fredholm alternative that equation (2.3) and hence (2.1) has a unique solution u_α in W for every $\alpha > 0$.

Because of the generally good behaviour of solutions to second kind Fredholm equations, we expect that (2.3) or equivalently (2.1) is a well posed analogue of the original improperly posed problem (1.1). This is indeed the case and as Tikhonov points out it depends on the following fundamental result.

<u>THEOREM 2.1</u> Let X be a Banach space with norm $\| \|_X$ and let $F : X \to L^2$ be 1-1 and continuous. If U is a compact subset of X , then the restriction of F to U , F/U , has a continuous inverse.

This means that given $\varepsilon > 0$ there exists $\delta > 0$ such that if $f_1, f_2 \in F(U)$ satisfy $\|f_1-f_2\|_2 \leq \delta$ then $\|u_1-u_2\|_X \leq \varepsilon$, where $u_1, u_2 \in U$ satisfy $F(u_1) = f_1$ and $F(u_2) = f_2$. Thus the restricted problem : For $f \in F(U)$ find $u \in U$ such that $Ku = f$, is well posed even though the unrestricted problem may be improperly posed.

Essentially the same thing occurs in regularization. If u_0 satisfies $Ku_0 = f$ then from (2.1)

$$\begin{aligned}\alpha\Omega(u_\alpha) &\leq \|Ku_\alpha-f\|_2^2 + \alpha\Omega(u_\alpha)\\ &\leq \|Ku_0-f\|_2^2 + \alpha\Omega(u_0)\\ &= \alpha\Omega(u_0)\ .\end{aligned}$$

Hence all regularized solutions u_α for $\alpha > 0$ belong to the set $U = \{u \in W : \Omega(u) \leq \Omega(u_0)\}$. Clearly this set is bounded in W. Let $X = C^{m-1}[0,1]$, the space of $m-1$ times continuously differentiable functions on $[0,1]$ with norm

$$\|u\|_X = \max_{0\leq i\leq m-1} \|u^{(i)}\|_\infty = \max_{0\leq i\leq m-1}\ \sup_{x\in[0,1]} |u^{(i)}(x)|\ .$$

Then there exists a compact imbedding $W \to X$, (see e.g. Adams [1]) which guarantees that the closure $\bar{U}$ of U is compact in X. Therefore in (2.3) we can restrict the domain of the 1-1 and continuous operator $GK^*K + \alpha I : X \to L^2$ to the compact set $\bar{U}$ and so apply Theorem 2.1. Because $GK^*K + \alpha I$ is linear, this implies that there exists a constant c such that

$$\|u_\alpha^{(i)}-\bar{u}_\alpha^{(i)}\|_\infty \leq c\|f-\bar{f}\|_2 \qquad i = 0,1,\ldots,m-1\ ,$$

where u_α and $\bar{u}_\alpha$ are the regularized solutions corresponding to functions f and $\bar{f}$ respectively. Thus Tikhonov regularization, considered as a map $C^{m-1}[0,1] \to L^2$, is well posed. It is possible to extend this result so that actually Tikhonov regularization is well posed as a map $W^{2m,2}(0,1) \to L^2$. The proof follows that of Theorem 3.4 of section 3.

Again from (2.1) it follows that

$$\begin{aligned}\|Ku_\alpha-f\|_2^2 &\leq \|Ku_\alpha-f\|_2^2 + \alpha\Omega(u_\alpha)\\ &\leq \|Ku_0-f\|_2^2 + \alpha\Omega(u_0)\\ &= \alpha\Omega(u_0)\\ &\to 0 \quad \text{as} \quad \alpha \to 0\ .\end{aligned}$$

Hence, if the null space of K, $N(K)$, is empty, then Theorem 2.1 with $F = K$ implies the following important convergence result

$$\|u_\alpha - u_0\|_X \le C\|Ku_\alpha - f\|_2 \to 0 \quad \text{as} \quad \alpha \to 0 .$$

That is u_α and its derivatives up to and including $m-1$ converge uniformly to u_0 and its corresponding derivatives as $\alpha \to 0$.

A similar result holds when there is error in f. If $\alpha = \alpha(\delta)$ converges to 0 as $0 < \delta \to 0$ in such a way that $0 < B \le \frac{\alpha(\delta)}{\delta^2} \le C$, then for every $\varepsilon > 0$ there exists a $\delta_0 > 0$ such that for all $\delta < \delta_0$ and $\bar{f} \in L^2$,

$$\|\bar{f} - f\|_2 \le \delta \quad \text{implies} \quad \|\bar{u}_{\alpha(\delta)} - u_0\|_X \le \varepsilon .$$

That is $\bar{u}_{\alpha(\delta)}$ and its derivatives up to and including $m-1$ converge uniformly to u_0 and its corresponding derivatives as $\delta \to 0$.

Note that if α tends to 0 independently of δ and $\|\bar{f}-f\|_2 = \delta$ then $\bar{u}_\alpha$ will not converge to u_0 in X.

3. Regularization with Differential Operators

For the original problem (1.1), another form of regularization when the norm and stabilizing functional are continuous is

$$\underset{u \in W}{\text{minimize}} \ \|Ku-f\|_2^2 + \alpha\|Tu\|_2^2 , \tag{3.1}$$

where $T : W \to L^2$ is a linear differential operator,

$$Tu = \sum_{i=0}^{m} w_i u^{(i)} \qquad w_i \in C^i[0,1] , \ \frac{1}{w_m} \in C[0,1] .$$

This is a generalization of Tikhonov regularization in the sense that if $\sum_{i=0}^{m} \int_0^1 w_i \left(u^{(i)}\right)^2 = 0$ for $w_i(x) > 0$ then $u = 0$, whereas $\Omega(u) = \|Tu\|_2^2 = 0$ for all u in the null space of T. Thus, the stabilizing functional can now be the square of a seminorm as well as a norm.

This is a very useful variation on Tikhonov regularization because in many problems a good measure of the smoothness of a trial solution u is its mean square slope $\|u'\|_2^2$ or linearized curvature $\|u''\|_2^2$; i.e. $Tu = u'$ or $Tu = u''$ above. In addition, the term $\int_0^1 w_0 u^2$ in the Tikhonov stabilizer has the effect of lowering the mean value of the regularized solution which is in general not desirable.

In Hilger's paper and thesis [11], [12] the regularized solution u_α of (3.1) is constructed when $Tu = u'$ and K is 1-1. The expression found is

$$u_\alpha = G(A^*A+\alpha I)^{-1}A^*(f-cb) + c ,$$

where G is the integral operator corresponding to the Green's function

$$G(x,y) = \begin{cases} 0 & x \le y \\ 1 & x > y \end{cases} ,$$

$$A = KG , \quad b(x) = K1(x) = \int_0^1 K(x,y)dy$$

and
$$c = \frac{((AA^*+\alpha I)^{-1}f,b)_2}{((AA^*+\alpha I)^{-1}b,b)_2} .$$

In practice this would be difficult to compute.

We now give a simpler constructive existence and uniqueness proof for general K and T. Let W be the space of smooth functions

$$\{u : u \in L^2 , \ u^{(m)} \text{ exists weakly and } Tu \in L^2\} ,$$

which is just the maximal domain of definition of

$$T : W \subseteq L^2 \to L^2 .$$

<u>THEOREM 3.1</u> If $T : W \to L^2$ is defined by

$$Tu = \sum_{i=0}^{m} w_i u^{(i)} \qquad w_i \in C^i[0,1] , \ \frac{1}{w_m} \in C[0,1] ,$$

then the problem

$$\underset{u\in W}{\text{minimize}} \ \|Ku-f\|_2^2 + \alpha\|Tu\|_2^2$$

has a solution, which is unique if $N(K) \cap N(T) = \{0\}$.

<u>Proof</u> A minimum will exist at u iff the function

$$M(\varepsilon,\eta) = \|K(u+\varepsilon\eta)-f\|_2^2 + \alpha\|T(u+\varepsilon\eta)\|_2^2$$

satisfies
$$\frac{d}{d\varepsilon}M(\varepsilon,\eta)\big|_0 = 0 \qquad \text{for all} \quad \eta \in W .$$

This is equivalent to

$$(Ku-f,K\eta)_2 + \alpha(Tu,T\eta)_2 = 0$$

or
$$(K^*Ku-K^*f,\eta)_2 = - \alpha(Tu,T\eta)_2 \qquad \text{for all} \quad \eta \in W ,$$

so that if u exists then it is contained in the domain of the operator T^*T and

$$(K^*K+\alpha T^*T)u = K^*f .$$

This is the same as

$$[(K^*K-\alpha I) + \alpha(T^*T+I)]u = K^*f . \tag{3.2}$$

Now because T is closed (Goldberg [10]) with dense domain W the operator $T^*T + I : \text{dom } T^*T \to L^2$ has a bounded inverse $G : L^2 \to \text{dom } T^*T$ which is symmetric and positive. Furthermore it is compact since it is an integral operator whose kernel is a Green's function. Hence (3.2) is equivalent to

$$G(K^*K-\alpha I)u + \alpha u = GK^*f . \tag{3.3}$$

This is a second kind Fredholm integral equation and therefore by the Fredholm alternative has a unique solution unless the homogeneous equation

$$G(K^*K-\alpha I)u + \alpha u = 0$$

has a non-zero solution. This is the same as

$$(K^*K-\alpha I)u + \alpha(T^*T+I)u = 0$$

or
$$K^*Ku = -\alpha T^*Tu$$

which implies
$$(Ku,Ku)_2 = -\alpha(Tu,Tu)_2 .$$

For $\alpha > 0$, this yields $Ku = 0$ and $Tu = 0$

or
$$u \in N(K) \cap N(T) .$$

If the homogeneous equation has a non-zero solution then (3.3) still has a (non-unique) solution if

$$(GK^*f,v)_2 = 0 \quad \text{for all} \quad v \in N([G(K^*K-\alpha I) + \alpha I]^*) .$$

But
$$(K^*K-\alpha I)Gv + \alpha v = 0 \quad \text{is equivalent to}$$

$$(K^*K-\alpha I)w + \alpha(T^*T+I)w = 0 \quad \text{and} \quad v = (T^*T+I)w .$$

As above, this implies

$$w \in N(K) \cap N(T) \quad \text{and} \quad v = (T^*T+I)w$$

or
$$v \in (T^*T+I)[N(K) \cap N(T)] .$$

Therefore $$(GK^*f,v)_2 = (f,KGv)_2 = 0 \ .$$

Thus the regularized solution has been constructed, which if $N(K) \cap N(T) = \{0\}$ can be written as

$$u_\alpha = [G(K^*K-\alpha I) + \alpha I]^{-1}GK^*f$$
$$= (K^*K+\alpha T^*T)^{-1}K^*f \ . \qquad ///$$

From the proof of Theorem 3.1, it is clear that the regularized solution u_α satisfies some natural boundary conditions defined by $u \in \text{dom } T^*T$. These are

$$D^j(Tu)(0) = \sum_{i=0}^{m} \left(w_i u^{(i)}\right)^{(j)}(0) = 0 \qquad j = 0,\ldots,m-1$$
$$\text{and} \qquad D^j(Tu)(1) = \sum_{i=0}^{m} \left(w_i u^{(i)}\right)^{(j)}(1) = 0 \qquad j = 0,\ldots,m-1 \ . \tag{3.4}$$

In a given problem, it may be desired to specify some different set of boundary conditions to these. For instance if $Tu = u'$ then (3.4) becomes

$$u'(0) = u'(1) = 0 \ .$$

If instead, it is required that the solution satisfy $u(0) = 0$, then this can be incorporated into the regularized solution using the following result.

<u>COROLLARY 3.2</u> Let $\{F_\theta\}$ be a set of continuous linear functionals on W, with the norm $\|u\|_W = \|u\|_2 + \|Tu\|_2$, for which $M = \bigcap_\theta F_\theta^{-1}(0)$ is dense in $L^2(0,1)$. If $T : M \to L^2$ is defined as before, then (3.1) has a solution in M, which is unique if $N(K) \cap N(T) = 0$.

<u>Proof</u> It suffices to show that $T : M \to L^2$ is closed, as then the proof will proceed exactly as for Theorem 3.1. If $u_n \in M$ and $u_n \to u$, $Tu_n \to f$ in L^2 for some $u,f \in L^2$, then, because $T : W \to L^2$ is closed, $u \in W$ and $Tu = f$. Also

$$\|u_n-u\|_W = \|u_n-u\|_2 + \|Tu_n-Tu\|_2$$
$$= \|u_n-u\|_2 + \|Tu_n-f\|_2$$
$$\to 0 \ .$$

Hence, because M is closed in W, $u \in M$. Thus $T : M \to L^2$ is closed. ///

Because $\|\cdot\|_W$ in Corollary 3.2 is equivalent to a Sobolev norm, it is not difficult to show that the boundary value functionals

$$B_i u = u^{(i)}(0) \qquad i = 0,1,\ldots,m-1$$

and $$B_i u = u^{(i)}(1) \qquad i = 0,1,\dots,m-1$$

are continuous. Hence any linear combination is also continuous. For the example above with $Tu = u'$ and $Bu = u(0) = 0$ specified, the condition $u \in \text{dom } T^*T$ is now equivalent to $u(0) = 0$ and $u'(1) = 0$.

A more general but less constructive existence and uniqueness result is proved in Ribiere [18]. Because Ribiere's development is directed mainly at obtaining convergence results and error estimates, the geometric nature of the existence theorem is a little obscured. For this reason, a very similar result is presented here more geometrically.

<u>THEOREM 3.3</u> Let $T : W \subseteq L^2 \to L^2$ be a closed linear operator which is onto and has finite dimensional null space $N(T)$, and let $K : L^2 \to L^2$ be a bounded linear operator. Then the problem

$$\underset{u \in W}{\text{minimize}} \ \|Ku-f\|_2^2 + \alpha\|Tu\|_2^2 \tag{3.5}$$

has a solution, which is unique if $N(K) \cap N(T) = \{0\}$.

<u>Proof</u> : Define a norm on $L^2 \oplus L^2$ by

$$\|g \oplus h\|^2 = \|g\|_2^2 + \alpha\|h\|_2^2 .$$

Then (3.5) is equivalent to the following :

$$\underset{u \in W}{\text{minimize}} \ \|Ku \oplus Tu - f \oplus 0\|^2 .$$

This is now a standard problem in Hilbert space which has a solution if the subspace $\{Ku \oplus Tu : u \in W\} \subseteq L^2 \oplus L^2$ is closed. The theorem follows by the lemma below.

<u>LEMMA 3.4</u> With the same assumptions as in the theorem, if for $u_n \in W$, $Ku_n \oplus Tu_n \to f \oplus g$ in $L^2 \oplus L^2$ then $f \oplus g = Ku \oplus Tu$ for some $u \in W$, which is unique if $N(K) \cap N(T) = \{0\}$.

<u>Proof</u> : Let $u_n = v_n + w_n$ where $v_n \in N(T)^\perp \cap W$, $w_n \in N(T)$. Then

$$\|Ku_n-f\|_2^2 + \alpha\|Tu_n-g\|_2^2 \to 0$$

implies $$\|Tu_n-g\|_2^2 = \|Tv_n-g\|_2^2 \to 0 .$$

By the open mapping theorem, there exists c such that

$$\|v_n-v_m\|_2 \le c\|Tv_n-Tv_m\|_2$$

so that $$\|v_n-v_m\|_2 \to 0 \quad \text{as} \quad n,m \to \infty .$$

Hence there exists a unique $v \in N(T)^{\perp} \subseteq L^2$ such that $v_n \to v$ in L^2. Now, because T is closed, $v \in \text{dom } T = W$ and $Tv = g$. Also, because K is continuous, $Kv_n \to Kv$ in L^2. In addition, since $Ku_n \to f$,

$$\|Kw_n - Kw_m\|_2 \le \|K[(w_n - w_m) + (v_n - v_m)]\|_2 + \|K(v_n - v_m)\|_2$$

$$= \|Ku_n - Ku_m\|_2 + \|Kv_n - Kv_m\|_2$$

$$\to 0 \quad \text{as} \quad n,m \to \infty .$$

Hence there exists a unique $h \in L^2$ such that $Kw_n \to h$. But, since $N(T)$ and so $K(N(T))$ is finite dimensional, $h \in K(N(T))$. That is $h = Kw$ for some $w \in N(T)$, which is unique if $K : N(T) \to L^2$ is 1-1 or equivalently $N(K) \cap N(T) = \{0\}$. Hence

$$Ku_n = Kv_n + Kw_n \to Kv + Kw = K(v+w)$$

with $$v \in N(T)^{\perp} \cap W \quad \text{and} \quad w \in N(T) .$$

Thus setting

$$u = v + w ,$$

we have $$Ku = Kv + Kw = f$$

and $$Tu = Tv + Tw = g .$$ ///

An example of the operator T considered above is the linear differential operator

$$Tu = \sum_{i=0}^{m} w_i u^{(i)} \qquad w_i \in C^i[0,1] , \quad \frac{1}{w_m} \in C[0,1]$$

defined earlier. Hence for this T and the integral operator K, Theorems 3.1 and 3.3 are equivalent.

In the remainder of this section assume that $N(K) \cap N(T) = \{0\}$. If $W = \text{dom } T$ is a Hilbert space, for instance $W = W^{m,2}(0,1)$, then Ribiere [18] proves the following continuity theorem. If u_α, $\bar{u}_\alpha$ are regularized solutions corresponding to f and $\bar{f}$, then there exists a constant c such that

$$\|u_\alpha - \bar{u}_\alpha\|_W \le c\|f - \bar{f}\|_2 .$$

Thus, there is a bound on the sensitivity of the regularized solution to perturbations in f. However, because $c \ge C\left(\frac{1}{\sqrt{\alpha}}\right)$ and the choice of α is typically $O(10^{-8})$ this bound is generally very large, expressing the fact that the problem (3.5) is almost improperly posed for small α.

It is possible to extend Ribiere's result to show that regularization with a differential operator is in fact well posed considered as a map $W^{2m,2}(0,1) \to L^2$.

THEOREM 3.4 If $N(K) \cap N(T) = \{0\}$ and u_α, $\bar{u}_\alpha$ are regularized solutions corresponding to f and $\bar{f}$, then there exists a constant c such that

$$\|u_\alpha - \bar{u}_\alpha\|_{W^{2m,2}(0,1)} \leq c\|f-\bar{f}\|_2 .$$

Proof : As in Theorem 3.1 the regularized solution u_α satisfies

$$(K^*K+\alpha T^*T)u = K^*f$$

or

$$(K^*K-\alpha I)u + \alpha(T^*T+I)u = K^*f .$$

Letting $v = (T^*T+I)u$ or equivalently $u = Gv$ this becomes

$$(K^*K-\alpha I)Gv + \alpha v = K^*f .$$

If $N(K) \cap N(T) = \{0\}$, this second kind Fredholm integral equation has a unique solution v_α. In addition, if $\bar{v}_\alpha$ is the solution corresponding to $\bar{f}$, then there exists a constant $c(\alpha)$ depending on α such that

$$\|v_\alpha - \bar{v}_\alpha\|_2 \leq c(\alpha)\|f-\bar{f}\|_2 .$$

Now since $G : L^2 \to W^{2m,2}(0,1)$ is bounded,

$$\begin{aligned} \|u_\alpha - \bar{u}_\alpha\|_{W^{2m,2}(0,1)} &= \|Gv_\alpha - G\bar{v}_\alpha\|_{W^{2m,2}(0,1)} \\ &\leq \|G\| \, \|v_\alpha - \bar{v}_\alpha\|_2 \\ &\leq \|G\| c(\alpha) \|f-\bar{f}\|_2 . \end{aligned}$$

///

The following important convergence theorem is proved by Ribiere. As $\alpha \to 0$, the regularized solution u_α converges in W to the unique solution $\hat{u}$ of $Ku = f$ which minimizes $\|Tu\|_2^2$.

Consider problem (3.5) in the equivalent form defined in Theorem 3.2,

i.e. $$\underset{u \in W}{\text{minimize}} \; \|Ku \oplus Tu - f \oplus 0\|^2 .$$

To solve it numerically, with $Tu = u^{(m)}$, Ribiere [18] discretizes K and T to obtain a discrete least squares problem, which he solves by a QR decomposition. By using splines as basis functions Locker and Prenter [16], obtain a continuous least squares problem for which a rigorous error analysis is developed.

Most of the results of this section were proved independently and in a different way by Locker and Prenter [15].

4. Discrete Regularization

On the basis of the preceding theory, there are several approaches that are possible for the numerical solution of the equation $Ku = f$. These can be grouped according to the stage at which discretization is introduced. That is

1. Functional minimization
2. Euler-Lagrange equation
3. Representation of the solution.

The first approach is to minimize a discrete form of the functional

$$\|Ku-f\|_2^2 + \alpha\Omega(u) . \tag{4.1}$$

We take $\Omega(u) = \|u'\|_2^2$ and assume that $N(K)$ contains no constant function. Then, from section 3, the continuous regularization problem has a unique solution u_α . Now consider a discrete version of this problem, namely,

$$\underset{\underset{\sim}{u}}{\text{minimize}} \sum_{i=0}^{n} \Big(\sum_{j=0}^{n} K_{ij}u_j\sigma_j - f_i\Big)^2\sigma_i + \alpha \sum_{i=0}^{n-1} \frac{(u_{i+1}-u_i)^2}{h} , \tag{4.2}$$

where $h = \frac{1}{n}$, $K_{ij} = K(ih,jh)$, $f_i = f(ih)$ and the σ_i are quadrature weights.

It can be shown that, for sufficiently small h , (4.2) has a unique solution $\underset{\sim}{u}_\alpha$ given by

$$\underset{\sim}{u}_\alpha = (K^tSKS + \tfrac{\alpha}{h}S^{-1}R)^{-1}K^tS\underset{\sim}{f}$$

where

$$S = \begin{bmatrix} \sigma_0 & & \\ & \ddots & \\ & & \sigma_n \end{bmatrix} \quad \text{and} \quad R = \begin{bmatrix} 1 & -1 & & & & \\ -1 & 2 & -1 & & & \\ & & \cdots & & & \\ & & & -1 & 2 & -1 \\ & & & & 1 & -1 \end{bmatrix} .$$

To show convergence of $\underset{\sim}{u}_\alpha$ to u_α in the sense $\sup_i |\underset{\sim}{u}_{\alpha_i} - u_\alpha(x_i)| \to 0$ as $h \to 0$ and to determine the rate of convergence, we compare $\underset{\sim}{u}_\alpha$ with the approximate solution derived by a second approach.

The unique minimizer of (4.1) is the unique solution of the Euler-Lagrange equation

$$K^*Ku - \alpha u'' = K^*f$$

with the boundary conditions $u'(0) = u'(1) = 0$. Discretizing this in a standard way on the extended grid $x_{-1} = -h$, $x_0,\dots,x_n$, $x_{n+1} = 1 + h$, we obtain

$$\sum_{j=0}^{n} (K^tSKS)_{ij}u_j - \frac{\alpha}{h^2}(u_{i-1}-2u_i+u_{i+1}) = \sum_{j=0}^{n} (K^tS)_{ij}f_j \quad i = 0,\dots,n , \tag{4.3}$$

$$u_{-1} = u_1 \qquad\qquad u_{n-1} = u_{n+1} .$$

In a slightly different form, this approach was first proposed by Tikhonov and Glasko [23], although they assumed the kernel $K(x,t)$ strictly positive on $[0,1] \times [0,1]$. We consider the general case of $K(x,t) \in L^2[(0,1) \times (0,1)]$ for which $\int_0^1 K(x,t)dt$ is not identically zero.

The existence and convergence of the solution of (4.3) can easily be derived using the stability theory of Linz [14] developed for general equations of the form $(L+\alpha A)u = f$. Sinc

$$A_n\underset{\sim}{u} = -\frac{(u_{i-1}-2u_i+u_{i+1})}{h^2}$$

with $\quad u_{-1} = u_1 \qquad u_{n-1} = u_{n+1}$

is a stable and consistent approximation to

$$Au = -u''$$

and $\quad u'(0) = u'(1) = 0$,

and if $\sup_n \sum_{i=1}^{n} |\sigma_i^{(n)}| < \infty$, then (4.3) is a stable and consistent approximation of

$$K^*Ku - \alpha u'' = K^*f$$

with boundary conditions $u'(0) = u'(1) = 0$.

This immediately implies that for sufficiently small h , (4.3) has a unique solution $\bar{\underset{\sim}{u}}_\alpha$ given by

$$\bar{\underset{\sim}{u}}_\alpha = (K^tSKS + \frac{\alpha}{h^2}P)^{-1}K^tS\underset{\sim}{f}$$

where

$$P = \begin{bmatrix} 2 & -2 & & & \\ -1 & 2 & -1 & & \\ & & \ddots & & \\ & & -1 & 2 & -1 \\ & & & -2 & 2 \end{bmatrix} .$$

In addition, $\bar{u}_{\alpha}$ converges to u_{α} in the sense that $\varepsilon_h = \sup_i |\bar{u}_{\alpha_i} - u_{\alpha}(x_i)| \to 0$ as $h \to 0$, and the discretization error ε_h is of the same order as the consistency error

$$\sup_i |K^*Ku(x_i) - K^tSKSu_i|$$

$$+ \alpha\left[\sup_i |u''(x_i) - \frac{1}{h^2}Pu_i| + \max\left(|u'(0) - \frac{u(x_1)-u(x_{-1})}{2h}| \ , \quad |u'(1) - \frac{u(x_{n+1})-u(x_{n-1})}{2h}|\right)\right]$$

Because the second term is $O(h^2)$, if a quadrature rule of $O(h^2)$ is used (for instance the midpoint rule), then for sufficiently small h, $\varepsilon_h \le C(\alpha)h^2$. The constant $C(\alpha)$ however will generally be very large, depending inversely on α. Therefore it would appear to be worthwhile to use a higher order rule, such as Simpson's rule which is of $O(h^4)$, so that if $h^2 \le \alpha$ then $\varepsilon_h \le C_1(\alpha)\alpha h^2$. However, in practice often $\alpha \ll h^2$ and then the error bounds become useless : $\varepsilon_h \le C(\alpha)h^2$ for the midpoint rule and $\varepsilon_h \le C_1(\alpha)h^4$ for Simpson's rule. Numerical experiments by de Hoog [13] indicate that for typical values of α and h, there is little dependence of $\bar{u}_{\alpha}$ on the quadrature rule used.

Returning to the first approach, it is clear that for the trapezoidal rule, the approximate solution is the same as that of the second approach, i.e. $u_{\alpha} = \bar{u}_{\alpha}$. Hence u_{α} converges $O(h^2)$ to u_{α}. For other quadrature rules only $O(h)$ convergence can be guaranteed. Evidently the reason for this is the influence of S^{-1} in the stabilizing term $\frac{\alpha}{h}S^{-1}R$ for u_{α}. The error bounds will again depend inversely on α.

For the general functional $\|Ku-f\|_2^2 + \alpha\|Tu\|_2^2$, where T is a linear differential operator with $N(T) \cap N(K) = \{0\}$, the regularized solution u_{α} is the unique solution of the Euler-Lagrange equation

$$(K^*K + \alpha T^*T)u = K^*f$$

with the natural boundary conditions. A quadrature and finite difference approximation to this leads to a similar linear system as above. In addition, similar arguments can be used to show existence, uniqueness and convergence of the approximate solution.

In both the first and second approach above the approximate solution is found to be increasingly unstable, when α decreases beyond a certain value, whereas close to or above this value the solution is relatively stable. This was observed by Tikhonov and Glasko [23] for the example they considered. An explanation can be given by considering an approximate solution derived by another approach.

With $\Omega(u) = \|u\|_2^2$ in (4.1) the regularized solution u_{α} is the unique solution of

$$(K^*K + \alpha I)u = K^*f \ . \tag{4.4}$$

Let $\{v_i\}$, $\{w_i\}$, $\{\lambda_i\}$ be a full system for K as defined in Smithies [20]. That is $\{v_i\}$ and $\{w_i\}$ are sets of orthonormal functions in L^2 which satisfy $K^*v_i = \lambda_i w_i$ and $Kw_i = \lambda_i v_i$. The λ_i satisfy $\lambda_0 \geq \lambda_1 \geq \ldots$ and $\lambda_n \to 0$ as $n \to \infty$. Then it is not hard to show that

$$u_\alpha = \sum_{i=0}^{\infty} \frac{\lambda_i}{\lambda_i^2+\alpha}(f,v_i)_2 w_i \; . \tag{4.5}$$

The true solution $u_0 \in L^2$ to $Ku = f$ when K is 1-1 and $f \in R(K)$ is

$$u_0 = \sum_{i=0}^{\infty} \frac{(f,v_i)_2}{\lambda_i} w_i \; .$$

Now, discretizing (4.4) for sufficiently small h, there is a unique solution $\underset{\sim}{u}_\alpha$ satisfying

$$(K^tSKS+\alpha I)\underset{\sim}{u} = K^tS\underset{\sim}{f} \; .$$

For simplicity, let $S = \mathrm{diag}(h,\ldots,h)$. Using a singular value decomposition, $\underset{\sim}{u}_\alpha$ can be written as

$$\underset{\sim}{u}_\alpha = \sum_{i=0}^{n} \frac{\mu_i h(\underset{\sim}{f},\underset{\sim}{v}_i)}{\mu_i^2+\alpha} \underset{\sim}{w}_i \; ,$$

where $(hK)^t\underset{\sim}{v}_i = \mu_i\underset{\sim}{w}_i$, $(hK)\underset{\sim}{w}_i = \mu_i\underset{\sim}{v}_i$ and $\mu_0 \geq \mu_1 \geq \ldots \geq \mu_n > 0$. Therefore, if α is very small,

$$\underset{\sim}{u}_\alpha \sim \sum_{i=0}^{n} \frac{h(\underset{\sim}{f},\underset{\sim}{v}_i)}{\mu_i} \underset{\sim}{w}_i$$

which is the same form as the true solution u_0. But for large $i \leq n$, μ_i will be very small so that errors resulting from the approximations $\lambda_i \to \mu_i$, $v_i \to \underset{\sim}{v}_i$, $w_i \to \underset{\sim}{w}_i$ and $(f,v_i)_2 \to h(\underset{\sim}{f},\underset{\sim}{v}_i)_{\mathbb{R}^{n+1}}$ will be greatly magnified. A similar instability occurs for general $\Omega(u) = \|Tu\|_2^2$ as was noted for $Tu = u'$ above.

For the third type of approach, the exact representation (4.5) of u_α is approximated directly. This requires estimation of the singular values λ_i and orthonormal eigenfunctions v_i of KK^* and w_i of K^*K (see e.g. Baker [7]). Note that the problem becomes simpler if K is self-adjoint, i.e. $K(x,t) = K(t,x)$.

Another representation of u_α, for a general linear differential operator T satisfying $N(T) \cap N(K) = \{0\}$, is

$$G(K^*K-\alpha I)u_\alpha + \alpha u_\alpha = GK^*f \; , \tag{4.6}$$

derived in section 3. Here G is the integral operator inverse of $T^*T + I$. For example if $Tu = u'$, then $Gu(x) = \int_0^1 G(x,t)u(t)$ where $G(x,t)$ is the Green's function for the boundary value problem

$$-u'' + u = 0$$

$$u'(0) = u'(1) = 0 .$$

i.e.
$$G(x,t) = C \begin{cases} \cosh x \cosh(t-1) & 0 \le x \le t \\ \cosh(x-1) \cosh t & t \le x \le 1 . \end{cases}$$

Equation (4.6) is a second kind Fredholm integral equation but because of the many operator products, it would not be practical to discretize and solve it directly.

If K is 1-1, self-adjoint, and positive then the following variation on regularization results in a simpler second kind Fredholm integral equation than (4.6). It is a generalization of a method given in Bakushinskii [8] where $T = I$. Since K is 1-1, self-adjoint, and positive there exists a 1-1, self-adjoint operator L such that $K = L^*L = L^2$. Let g be defined by $Lg = f$. Applying regularization to the equation $Lv = g$ defines a unique stabilized solution v_α of $Kv = f$, by the equation

$$(L^*L+\alpha T^*T)v = L^*g ,$$

i.e.
$$(K+\alpha T^*T)v = f .$$

This equation could be discretized directly or transformed to the second kind Fredholm integral equation

$$G(K-\alpha I)v + \alpha v = Gf$$

and then solved numerically by known methods.

5. Regularization with Discrete Data

We now examine in some detail the situation when the function f in (1.1) is only known approximately at discrete points; namely,

$$d_i = f(x_i) + \varepsilon_i \qquad i = 1,\dots,n ,$$

where ε_i are random errors. This situation occurs regularly in applications.

For this and the next section the relevant variational formulation is :

$$\underset{u \in W}{\text{minimize}} \sum_{i=1}^{n} w_i (Ku(x_i)-d_i)^2 + \alpha \|Tu\|_2^2 , \tag{5.1}$$

where $T : W \to L^2$ is a linear operator. This can be interpreted as a discretization of the first term in (3.1). Without loss of generality we shall take $w_i \equiv 1$.

Before treating (5.1) we consider the curve fitting analogue defined with $K = I$ and $T = D^m$, $m \geq 0$; namely,

$$\underset{u \in W}{\text{minimise}} \sum_{i=1}^{n} (u(x_i)-d_i)^2 + \alpha \|D^m u\|_2^2 , \tag{5.2}$$

with $W = W^{m,2}(0,1)$. This smoothing problem is fundamentally related to the interpolation problem

$$\underset{u \in W}{\text{minimize}} \|D^m u\|_2^2 \quad \text{subject to} \quad u(x_i) = d_i , \quad i = 1,\dots,n . \tag{5.3}$$

For instance, a simple argument by contradiction shows that if $u = s(x;\{x_i,d_i\})$ solves the interpolation problem then $v = s(x;\{x_i,d_i\})$ solves the smoothing problem for some d_i . Thus the two solutions, if they are unique, will have the same form.

The fundamental result for (5.2) (due to Schoenberg [19]) and for (5.3) (due to de Boor and Lynch [9]) is the following.

THEOREM 5.1 Both the interpolation and smoothing problems have unique solutions which are natural splines of degree $2m - 1$.

A natural spline of degree $2m - 1$, s_{2m-1} , on the interval $[0,1]$ with knots at x_i , such that $0 < x_1 < \dots < x_n < 1$, is defined by

(a) $s_{2m-1} \in P_{2m-1}$ on (x_i,x_{i+1}) $i = 1,\dots,n-1$

(b) $s_{2m-1} \in P_{m-1}$ on $(0,x_1)$ and $(x_n,1)$

(c) $s_{2m-1} \in C^{2m-2}[0,1]$.

Here P_k denotes the space of polynomials of degree $\leq k$.

It is not difficult to see that the set of all natural splines of degree $2m-1$, N_{2m-1} , is a vector space of dimension n . To construct a spanning set for N_{2m-1} take the polynomials x^j $j = 0,\dots,m-1$ and the truncated polynomials $(x-x_i)_+^{2m-1}$ $i = 1,\dots,n$, where

$$(x-x_i)_+^{2m-1} = \begin{cases} 0 & x - x_i \leq 0 \\ (x-x_i)^{2m-1} & x - x_i \geq 0 . \end{cases}$$

Thus any $s \in N_{2m-1}$ can be written as

$$s(x) = a_0 + a_1 x + \ldots + a_{m-1}x^{m-1} + \sum_{i=1}^{n} b_i (x-x_i)_+^{2m-1} .$$

An important set of basic spline functions consists of the so called B-splines. To define these, we first define the m'th divided difference of a function f with respect to knots $\{x_j \; ; \; j = i,\ldots,i+m\}$, inductively by

$$f[x_i] = f(x_i) \qquad i = 1,\ldots,n$$

$$f[x_i,x_{i+1}] = \frac{f[x_{i+1}]-f[x_i]}{x_{i+1}-x_i} \qquad i = 1,\ldots,n-1$$

$$\cdots$$

$$f[x_i,\ldots,x_{i+m}] = \frac{f[x_i,\ldots,x_{i+m}]-f[x_i,\ldots,x_{i+m-1}]}{x_{i+m}-x_i} \qquad i = 1,\ldots,n-m .$$

A B-spline of order m is then defined to be the m'th divided difference, where $m \geq 2$, of

$$M_m(x;t) = m(t-x)_+^{m-1}$$

with respect to the knots $t = x_j \quad j = i,\ldots,i+m$ and is denoted by $M_{mi}(x) = M_m[x;x_i,\ldots,x_{i+m}]$. For all $m \geq 2$, M_{mi} is a piecewise polynomial of degree $m - 1$ which is in $C^{m-2}(\mathbb{R})$ and has compact support equal to $[x_i,x_{i+m}]$. For $m = 2$ the B-splines are the linear hat functions.

The importance of B-splines lies in the fact that the set $\{M_{mi} : i = 1,\ldots,n-m\}$ forms a basis for the space $D^m(N_{2m-1})$. From above this is the same as $D^m(sp\{(x-x_i)_+^{2m-1} : i = 1,\ldots,n\})$. This provides the key to the method of Anselone and Laurent [5] which we now describe in greater generality.

To continue in the right framework we make the following definition.

DEFINITION 5.1 A reproducing kernel Hilbert space (RKHS) W is a Hilbert space of real valued functions on $[0,1]$, in which, for each $x \in [0,1]$, there exists $R_x \in W$ such that $(R_x,f)_W = f(x)$ for all $f \in W$.

Because $(R_x,R_y)_W = R_y(x) = R_x(y)$, the kernel $R(x,y) = R_x(y)$ is called the reproducing kernel (RK) of W . For properties of these spaces see Aronszajn [6].

By the Riesz representation theorem, an equivalent definition of a RKHS is a Hilbert space of functions $f : [0,1] \to \mathbb{R}$ in which, for each $x \in [0,1]$, the linear evaluation functional $N_x : W \to \mathbb{R}$, $N_x f = f(x)$, is bounded. Note that L^2 is not a RKHS. If $L : W \to \mathbb{R}$ is a bounded linear functional on a RKHS W then the unique representer ψ satisfying $Lf = (\psi,f)_W$ for all $f \in W$, is given by $\psi(x) = LR_x$.

Let $W = W^{m,2}(0,1)$ and let $T : W \to L^2$ be continuous with null space of dimension m. By showing that all the evaluation functionals $f \to f(x)$ are bounded, it is not difficult to prove that W is a RKHS. In addition, because

$$\left|\int_0^1 K(x,t)u(t)dt\right| \le \|K(x,\cdot)\|_2\|u\|_2$$
$$\le C\|K(x,\cdot)\|_2\|u\|_W$$

and $\|K(x,\cdot)\|_2 < \infty$ by Fubini's theorem, the linear functionals

$$K_x : W \to \mathbb{R}, \qquad K_x u = \int_0^1 K(x,t)u(t)dt = Ku(x)$$

are bounded. Hence their representers k_x are given by $k_x(t) = K_x R_t$, where R is the RK for W. The special case of $K = I$ considered earlier also fits into this framework since the functionals $W \to \mathbb{R}$, $u \to Iu(x) = u(x)$ are bounded, being just the evaluation functionals.

Returning to the variational formulation (5.1), it can now be written as

$$\underset{u \in W}{\text{minimize}} \sum_{i=1}^{n} ((k_i,u)_W - d_i)^2 + \alpha\|Tu\|_2^2 , \tag{5.4}$$

where $k_i = k_{x_i}$. This illustrates the advantage of the RKHS framework, as (5.4) is now a geometrical problem in a Hilbert space. The same is true for the corresponding interpolation problem, which now becomes

$$\underset{u \in W}{\text{minimize}} \|Tu\|_2^2 \quad \text{subject to} \quad (k_i,u)_W = d_i , \quad i = 1,\ldots,n . \tag{5.5}$$

As before, these two problems are closely related and following Anselone and Laurent [5] we treat the interpolation problem first.

Assume $\{k_1,\ldots,k_n\}$ is linearly independent, for otherwise the problem may have no solution. Set

$$\kappa = \text{sp}\{k_1,\ldots,k_n\}$$

and

$$\kappa_d^\perp = \{u \in W : (k_i,u)_W = d_i , \quad i = 1,\ldots,n\} .$$

It is easy to see that $\kappa_d^\perp$ is just a translate of $\kappa^\perp$. We now state the following basic result.

<u>THEOREM 5.2</u> If $T(\kappa^\perp)$ is closed and $N(T) \cap \kappa^\perp = \{0\}$ then for each $d \in \mathbb{R}^n$ there exists a unique $\sigma \in W$ which solves the interpolation problem (5.5). It is determined by $\sigma \in \kappa_d^\perp$ and $T\sigma \in (T(\kappa^\perp))^\perp$.

It can be shown that $T(\kappa^\perp)$ is closed whenever $N(T) \cap \kappa^\perp = \{0\}$ and $R(T)$ is closed. We will assume that $R(T) = L^2$ which is closed.

Since the conditions above determining σ are non-constructive, the aim is to construct a basis $\{f_i\}$ for $F = (T(\kappa^\perp))^\perp$ so that $T\sigma = \Sigma\lambda_i f_i$. This can be done by showing that F has the alternative form $F = T^{*-1}(N(T)^\perp \cap \kappa)$. It has dimension $n - m$ because $N(T)^\perp \cap \kappa$ has, and T^{*-1} exists since $N(T^*) = R(T)^\perp = \{0\}$. Let $\{b_i = (b_i^1,\dots,b_i^n) : i = 1,\dots,n-m\}$ be a linearly independent set in $\mathbb{R}^n$ satisfying $\sum_{j=1}^n b_i^j(k_j,\varphi)_W = 0$ for all $\varphi \in N(T)$ and $i = 1,\dots,n-m$. Then define the basis $\{f_i\}$ for F by $T^*f_i = \sum_{j=1}^n b_i^j k_j$. In general, however, this construction will be difficult.

Now, from above, we have $T\sigma = \sum_{i=1}^{n-m} \lambda_i f_i$ and the λ_i satisfy the linear system

$$\sum_{i=1}^{n-m} \lambda_i (f_i,f_\ell)_2 = (b_\ell,d)_{\mathbb{R}^n} = \sum_{j=1}^n b_\ell^j d_j \qquad \ell = 1,\dots,n-m .$$

To determine σ, solve $T\sigma = \sum_{i=1}^{n-m} \lambda_i f_i$ with the conditions associated with $\sigma \in \kappa_d^\perp$ i.e. $(k_j,\sigma) = d_j$, $j = 1,\dots,n$. In practice this may be quite difficult.

We now illustrate this method for the case when $K = I$ and $T = D^m$. Here there exist $k_j \in W$ which satisfy $(k_j,u)_W = u(x_j)$ for all $u \in W$. The b_i above can be constructed from m'th divided differences by taking

$$b_i^j = \text{coefficient of } f(x_i) \text{ in } f[x_i,\dots,x_{i+m}] \qquad i = 1,\dots,n-m$$

$$= \begin{cases} 0 & j = 1,2,\dots,i-1,i+m+1,\dots,n \\ [(x_j-x_i)\dots(x_j-x_{j-1})(x_j-x_{j+1})\dots(x_j-x_{i+m})]^{-1} & j = i,\dots,i+m . \end{cases}$$

Then for each $i = 1,\dots,n-m$,

$$\sum_{j=1}^n b_i^j(k_j,\varphi)_W = \sum_{j=1}^n b_i^j\varphi(x_j) = 0$$

for all polynomials φ of degree $\le m-1$, i.e. $\varphi \in N(T)$. Let $h_i = \sum_{j=1}^{n} b_i^j k_j$. To construct the basis $\{f_i : i = 1,\dots,n-m\}$ first note that by the Peano kernel theorem there exist functions $\psi_i \in W$ such that

$$u[x_i,\dots,x_{i+m}] = \int_0^1 \psi_i u^{(m)} .$$

Thus

$$u[x_i,\dots,x_{i+m}] = (h_i,u)_W = (\psi_i,u^{(m)})_2 = (\psi_i,Tu)_2$$

and therefore $h_i = T^* \psi_i$. Hence $\{f_i = \psi_i : i = 1,\dots,n-m\}$ is a basis for F and in fact the ψ_i are just the B-splines discussed earlier. We have $T\sigma = \sum_{i=1}^{n-m} \lambda_i \psi_i$, and the matrix $[(\psi_i,\psi_\ell)_2]$ that must be inverted to calculate the λ_i's is sparse, with $(\psi_i,\psi_\ell)_2 = 0$ for $|i-\ell| \ge m$. Finally for σ , we integrate m times and use the conditions $\sigma(x_j) = d_j$ to obtain the natural spline of Theorem 5.1.

For the general smoothing problem (5.4), the method of solution is very similar to that of the interpolation problem above. Define a bounded linear map L by

$$L : W \to \mathbb{R}^n \times L^2 , \qquad Lu = [((k_1,u)_W,\dots,(k_n,u)_W);Tu] .$$

Then (5.4) is equivalent to

$$\underset{u \in W}{\text{minimize}} \ \|Lu - [d;0]\|^2_{\mathbb{R}^n \times L^2} ,$$

where the inner product on $\mathbb{R}^n \times L^2$ is

$$([r;v],[p;w])_{\mathbb{R}^n \times L^2} = (r,p)_{\mathbb{R}^n} + \alpha(v,w)_2 .$$

From this follows another basic result of Anselone and Laurent [5].

<u>THEOREM 5.3</u> If $L(W)$ is closed in $\mathbb{R}^n \times L^2$ and $N(T) \cap K^\perp = \{0\}$ then there exists a unique $s \in W$ which solves the smoothing problem (5.1) It is determined by the condition $Ls - [d;0] \in L(W)^\perp$.

It can be shown that $L(W)$ is closed whenever $N(T) \cap K^\perp = \{0\}$ and $R(T)$ is closed. Again we assume that $R(T) = L^2$ which is closed.

It is not difficult to show that $L(W)$ has finite dimension $n-m$ and a basis for it is $\{g_i = [f_i;b_i]\}$, where f_i and b_i have been defined above. Then we have

$$Ls - [d;0] = \sum_{i=1}^{n-m} \mu_i g_i \ ,$$

where the μ_i satisfy the linear system

$$\sum_{i=1}^{n-m} \mu_i (g_i, g_\ell)_{\mathbb{R}^n \times L^2} = ([d;0], g_\ell)_{\mathbb{R}^n \times L^2} \qquad \ell = 1,\ldots,n-m \ .$$

That is

$$Ts = \sum_{i=1}^{n-m} \mu_i f_i$$

and

$$((k_1,s)_W,\ldots,(k_n,s)_W) = d - \sum_{i=1}^{n-m} \mu_i b_i \ .$$

Hence the form of s is exactly the same as that of σ , the difference between them simply being that they correspond to different data.

For the case of $K = I$ and $T = D^m$ we have $Ts = \sum_{i=1}^{n-m} \mu_i \psi_i$ where ψ_i are the B-splines considered before. The matrix $(g_i, g_\ell)_{\mathbb{R}^n \times L^2}$ is again sparse. On integrating to obtain s , the conditions

$$s(x_j) = d_j - \sum_{i=1}^{n-m} \mu_i b_i^j \qquad j = 1,\ldots,n$$

must be used.

6. An Implementation with Differential Operators

In this section, an implementation using differential operators of the regularization method of section 5 is given, by a different approach due to Wahba [24]. A comparison is made between this and the approach in section 5 due to Anselone and Laurent [5].

Let T be an m'th order linear differential operator

$$T = \sum_{i=1}^{m} w_i D^i : W^{m,2} \to L^2 \qquad w_i \in C^i[0,1] \ , \ \frac{1}{w_m} \in C[0,1]$$

and let $\{\varphi_i : i = 1,\ldots,m\}$ be a basis for the null-space of T . Consider a set of linear homogeneous boundary conditions $B_i u = 0 \quad i = 1,\ldots,m$ for which

$$Tu = 0 \ , \qquad B_i u = 0 \qquad i = 1,\ldots,m$$

has only the zero solution. Then the Green's function $G(x,t)$ exists and the operator G given by

$$Gu(x) = \int_0^1 G(x,t)u(t)dt, \quad G : L^2 \to \{u : u \in W^{m,2} , B_i u = 0\}$$

satisfies $TG = I$. This situation is a generalization of the initial conditions

$$M_i u = \sum_{j=1}^{m} a_{ij} u^{(j)}(0)$$

considered by Wahba [24], but the analysis proceeds in much the same way. Note that the boundary conditions $B_i u = 0$ will not be imposed on the regularized solution.

Because the matrix $[B_i \varphi_j]$ is non-singular, the basis elements φ_i can be chosen so that $B_i \varphi_j = \delta_{ij}$. Define an inner product on $N(T)$ by

$$(u,v)_{N(T)} = \sum_{i=1}^{m} B_i u B_i v \ .$$

Then $N(T)$ is a RKHS with RK

$$R_0(x,y) = \sum_{i=1}^{m} \varphi_i(x)\varphi_i(y) \ .$$

The range of G with the inner product

$$(u,v)_{R(G)} = (Tu,Tv)_2$$

is also a RKHS with RK

$$R_1(x,y) = GG^*(x,y) = \int_0^1 G(x,s)G(y,s)ds \ .$$

Because $W^{m,2} = N(T) \oplus R(G)$, define W to be $W^{m,2}$ with the inner product

$$(u,v)_W = (P_0 u, P_0 v)_{N(T)} + (P_1 u, P_1 v)_{R(G)}$$

where $P_1 = GT$ is a projection onto $R(G)$ and $P_0 = I - GT$ is a projection onto $N(T)$. Then W is a RKHS with RK $R = R_0 + R_1$. Although $W^{m,2}$ with the Sobolev inner product is also a RKHS , it is not used here because an explicit form for its RK is not known.

<u>LEMMA 6.1</u> The linear functionals $K_x : W \to \mathbb{R}$, $x \in [0,1]$,

$$K_x u = \int_0^1 K(x,t)u(t)dt$$

are bounded.

Proof Let $u = u_0 + u_1$ where $u_0 \in N(T)$ and $u_1 \in R(G)$. Then

$$\begin{aligned} |K_x u| &= \left|\int_0^1 K(x,t)(u_0(t)+u_1(t))dt\right| \\ &\le \|K(x,\cdot)\|_2\|u_0\|_2 + \|K(x,\cdot)\|_2\|GTu_1\|_2 \\ &\le C_0\|K(x,\cdot)\|_2\|u_0\|_{N(T)} + \|K(x,\cdot)\|_2\|G\|\|Tu_1\|_2 \\ &\le C\|u\|_W . \end{aligned}$$

The second last step follows because norms on a finite dimensional space are equivalent and because $G : L^2 \to L^2$ is bounded. ///

Note that Lemma 6.1 could also be proved by showing that the norms on W and $W^{m,2}$ are equivalent.

From Lemma 6.1 and the Riesz representation theorem, there exist representers $k_x \in W$ such that

$$K_x u = (k_x,u)_W .$$

These k_x are known explicitly; namely

$$\begin{aligned} k_x(t) &= K_x(R_t) \\ &= \sum_{i=1}^m \varphi_i(t)K\varphi_i(x) + KGG^*(x,t) . \end{aligned}$$

At the points x_i $i = 1,\dots,n$, we define $k_i(t) = k_{x_i}(t)$.

Because $\|Tu\|_2^2 = \|P_1u\|_W^2$, the problem (5.1) is equivalent to

$$\underset{u\in W}{\text{minimize}} \sum_{i=1}^n ((k_i,u)_W-d_i)^2 + \alpha\|P_1u\|_W^2 . \tag{6.1}$$

This is now a standard Hilbert space projection problem. It has a unique solution which can be constructed as follows.

Let the vectors $\underset{\sim}{\varphi}$ and $\underset{\sim}{\xi}$ be defined by

$$\underset{\sim}{\varphi}_i = \varphi_i \qquad i = 1,\dots,m$$

$$\underset{\sim}{\xi}_i = P_1 k_i = KGG^*(x_i,\cdot) \qquad i = 1,\dots,n$$

$$= \int_0^1 K(x_i,s)GG^*(s,\cdot)ds ,$$

and the matrices X , Σ and Σ_α by

$$X_{ij} = (\varphi_i,k_j)_W = \int_0^1 K(x_j,s)\varphi_i(s)ds \qquad i = 1,\dots,m , \quad j = 1,\dots,n$$

$$\Sigma_{ij} = (\underset{\sim}{\xi}_i,\underset{\sim}{\xi}_j)_W = \int_0^1 \int_0^1 K(x_i,s)GG^*(s,t)K(x_j,t)dsdt \qquad i,j = 1,\dots,n$$

$$\Sigma_\alpha = \Sigma + \alpha I .$$

Then the unique solution of (6.1) is given by

$$u_\alpha = \underset{\sim}{\varphi}(X\Sigma_\alpha^{-1}X^t)^{-1}X\Sigma_\alpha^{-1}\underset{\sim}{d}^t + \underset{\sim}{\xi}(\Sigma_\alpha^{-1}-\Sigma_\alpha^{-1}X^t(X\Sigma_\alpha^{-1}X^t)^{-1}X\Sigma_\alpha^{-1})\underset{\sim}{d}^t . \qquad (6.2)$$

As above, the interpolation problem (5.5) is equivalent to

$$\underset{u\in W}{\text{minimize}} \; \|P_1 u\|_W^2 \quad \text{subject to} \quad (k_i,u)_W = d_i \qquad i = 1,\dots,n .$$

This has the unique solution

$$u_0 = \underset{\sim}{\varphi}(X\Sigma^{-1}X^t)^{-1}X\Sigma^{-1}\underset{\sim}{d}^t + \underset{\sim}{\xi}(\Sigma^{-1}-\Sigma^{-1}X^t(X\Sigma^{-1}X^t)^{-1}X\Sigma^{-1})\underset{\sim}{d}^t . \qquad (6.3)$$

From (6.2) and (6.3) it follows that $u_\alpha \to u_0$ in W as $\alpha \to 0$.

This shows how the smoothing and interpolation problems are related. Both solutions are finite linear combinations of the functions φ_i (which depend on T) and ξ_i (which depend on T , K and grid points x_i). This property, that the regularized solution is restricted to a finite dimensional subspace independent of α and the data, is a feature of this form of regularization. Compare this with the infinite dimensional space of possible regularized solutions obtained by varying the function f in Tikhonov regularization or regularization with a differential operator.

When $K = I$ and $T = D^m$, the problems of data smoothing (5.2) and interpolation (5.3) also fit into the above framework, since the linear evaluation functionals

$$K_{x_i} : W \to \mathbb{R} , \quad K_{x_i}u = u(x_i) \qquad i = 1,\dots,n$$

are bounded. Their representers are

$$k_i(t) = R(x_i,t) = \sum_{j=1}^{m} \varphi_j(x_i)\varphi_j(t) + GG^*(x_i,t) \ .$$

Hence the solutions to (5.2) and (5.3) both have the form

$$\sum_{i=1}^{m} a_i\varphi_i + \sum_{i=1}^{n} b_i GG^*(x_i,\cdot) \ , \tag{6.4}$$

where the a_i , b_i can be computed as in (6.2) and (6.3). Here $\{\varphi_i\}$ is a basis for the space of polynomials of degree $\leq m - 1$ and, for each i , $GG^*(x_i,\cdot) \in C^{2m-2}[0,1]$ is a polynomial of degree $2m - 1$ on each of the intervals $[0,x_i]$ and $[x_i,1]$. In fact, (6.4) is the natural spline solution of section 5.

The above approach and the one in section 5 can be related, (Lukas [17]). For the Anselone and Laurent formulation, it can be shown that the solution of the smoothing problem (5.1) is

$$u_\alpha = \sum_{i=1}^{m} c_i\varphi_i + \underset{\sim}{\xi} B^t(B\Sigma_\alpha B^t)^{-1}B\underset{\sim}{d}^t \ , \tag{6.5}$$

where

$$X^t\underset{\sim}{c} = \underset{\sim}{d}^t - \Sigma_\alpha B^t(B\Sigma_\alpha B^t)^{-1}B\underset{\sim}{d}^t \ ;$$

and that of the interpolation problem (5.5) is

$$u_0 = \sum_{i=1}^{m} a_i\varphi_i + \underset{\sim}{\xi} B^t(B\Sigma B^t)^{-1}B\underset{\sim}{d}^t \ , \tag{6.6}$$

where

$$X^t\underset{\sim}{a} = \underset{\sim}{d}^t - \Sigma B^t(B\Sigma B^t)^{-1}B\underset{\sim}{d}^t \ .$$

Here $B = [b_i^j]$ $i = 1,\ldots,n-m$ $j = 1,\ldots,n$ with the b_i^j the coefficients defined in section 5. The $n - m$ row vectors b_i are linearly independent and satisfy

$$\left(\sum_{j=1}^{n} b_i^j k_j,\varphi\right)_W = 0$$

for all $\varphi \in N(T)$ and $i = 1,\ldots,n-m$. This is the same as

$$\sum_{j=1}^{n} b_i^j (k_j,\varphi_\ell)_W = 0 \qquad i = 1,\ldots,n-m \quad \ell = 1,\ldots,m \ ;$$

namely

$$BX^t = 0 \ .$$

With this condition on B it is not difficult to show that the expressions (6.2) and (6.3) are equivalent to (6.5) and (6.6) respectively.

Although, for the cases they considered, Anselone and Laurent were able to compute the b_i from divided differences, in general this will not be possible. Instead, the b_i will have to be found as a basis for the orthogonal complement in $\mathbb{R}^n$ of the span of the row vectors of X. This means that $B\Sigma B^t$ and $B\Sigma_\alpha B^t$ (equal to $[(f_i,f_j)_2]$ and $[(g_i,g_j)_{\mathbb{R}^n \times L^2}]$ in section 5) will not, in general, be easily made diagonally dominant. Hence if $m \ll n$, the expressions (6.2) and (6.3) would be more efficient in evaluating the solutions than (6.5) and (6.6), since the matrix products are of smaller size.

7. Weighted Cross-Validation for the Choice of α.

In the above discussion of different formulations for regularization and the construction of regularized solutions u_α, no mention has been made about the choice of α. In practical situations, when the data contains significant errors, the choice of α is critical. Too small an α will yield a poor approximate solution u_α, containing greatly magnified errors, while too large an α will yield too smooth an approximation.

In addition, the "optimal" α and corresponding u_α can depend quite crucially on the actual errors present in the data rather than just their statistical properties. This was observed in numerical experiments by de Hoog [13]. It is therefore essential that any method for choosing α be based on the data being analysed.

One such method is weighted cross-validation of Wahba [25]. Given the equation $Ku = f$ with discrete data

$$d_i = f(x_i) + \varepsilon_i \qquad i = 1,\ldots,n ,$$

consider the regularization problem

$$\underset{u\in W}{\text{minimize}}\ \frac{1}{n}\sum_{i=1}^{n} (Ku(x_i) - d_i)^2 + \alpha\|u\|_W^2 , \tag{7.1}$$

where W is either L^2 or a RKHS with RK $R(x,y)$. As in sections 5 and 6, it can be shown that (7.1) has a unique solution u_α.

Following Wahba we assume that f is "very smooth". This is defined in terms of the kernel

$$Q(x,y) = \begin{cases} \int_0^1 K(x,t)K(y,t)\,dt & \text{if } W = L^2 \\ \int_0^1\int_0^1 K(x,s)K(y,t)R(s,t)\,ds\,dt & \text{if } W \text{ is a RKHS}, \end{cases}$$

which has a Hilbert Schmidt expansion

$$Q(x,y) = \sum \lambda_i \varphi_i(x) \varphi_i(y) \qquad \lambda_i \neq 0 \; .$$

Then, we say that f is "very smooth" if

$$|||f|||^2 = \sum \frac{|(f,\varphi_i)_2|^2}{\lambda_i^2} < \infty \quad .$$

For f in the span of $\{\varphi_i\}$, this implies that f is in the range of Q , where $Q : L^2 \to L^2$ is defined by

$$Qu(x) = \int_0^1 Q(x,y)u(y)dy \; .$$

If the kernel $K(x,t)$ is smooth, then the kernel $Q(x,y)$ will be even smoother so that this is quite a strong smoothness requirement for f . The errors ε_i are assumed to be uncorrelated zero mean random variables with common variance σ^2 .

The essence of weighted cross-validation is the following. Let $u_{\alpha,k}$ be the minimizer of

$$\frac{1}{n} \sum_{i \neq k} (Ku(x_i)-d_i)^2 + \alpha \|u\|_W^2 \; ,$$

where the k'th data point has been dropped. If $\hat{\alpha}$ is a good choice of the regularization parameter then for each k , $Ku_{\hat{\alpha},k}(x_k)$ should be closer to d_k on average than $Ku_{\alpha,k}(x_k)$ for other values of α . Thus, we choose $\hat{\alpha}$ on the basis of the remaining data's ability to predict, for each k , the missing k'th data value. A measure of this is the mean square prediction error

$$V(\alpha) = \frac{1}{n} \sum_{k=1}^{n} (Ku_{\alpha,k}(x_k)-d_k)^2 w_k(\alpha) \; .$$

The weights $w_k(\alpha)$ are chosen so that $w_k(\alpha)$ will emphasize the k'th prediction error when it has small variance. The weighted cross-validation choice for α is then the minimizer $\hat{\alpha}$ of $V(\alpha)$.

Let $T(\alpha)$ denote the mean square true prediction error; namely,

$$T(\alpha) = \frac{1}{n} \sum_{i=1}^{n} (Ku_\alpha(x_i)-f(x_i))^2 \; .$$

For $\hat{\alpha}$ to be a good choice of α , it should be close to the minimizer of $T(\alpha)$. We now assume that the eigenvalues of Q , $\{\lambda_i\}$, decay like $\lambda_i = 1/C_i i^{2m}$ with $0 < B_1 \leq 1/C_i \leq B_2 < \infty$ and $m \geq 1$. Also assume that $\{x_i : i = 1,\ldots,n\}$ is an even grid and n is sufficiently large. Under such conditions Wahba shows that if α^* and $\tilde{\alpha}$ minimize $EV(\alpha)$ and $ET(\alpha)$ respectively, then

$$\alpha^* = \tilde{\alpha}(1+o(1)) ,$$

where $o(1) \to 0$ as $n \to \infty$, and

$$\tilde{\alpha} = O\left(\left[\frac{k_m \sigma^2}{4m \, |||f|||^2 n}\right]^{\frac{2m}{4m+1}}\right)$$

where k_m is a constant depending on m.

As it stands the above formula for $V(\alpha)$ is not suitable for the computation of $\hat{\alpha}$, since it requires the evaluation of $u_{\alpha,k}$ and then $Ku_{\alpha,k}(x_k)$ for each k and many values of α. A computationally efficient procedure is obtained if the alternative form

$$V(\alpha) = \frac{1}{n} \sum_{i=1}^{n} \left(\frac{\hat{d}_i}{\mu_i + n\alpha}\right)^2 \Bigg/ \left[\frac{1}{n} \sum_{i=1}^{n} \frac{1}{(\mu_i + n\alpha)}\right]^2 ,$$

derived by Wahba, is used. Here, the μ_i $i = 1,\ldots,n$ are the eigenvalues of the matrix $[Q(x_i,x_j)]$ and if φ_i $i = 1,\ldots,n$ are corresponding orthonormal eigenvectors then

$$\hat{d}_i = \sum_{j=1}^{n} (\varphi_i)_j d_j .$$

If $Q(s,t)$ is of the form $q(s-t)$ where q is periodic with period 1, then an approximation to $V(\alpha)$ can be easily obtained. In fact

$$Q(s,t) = \sum_{k=-\infty}^{\infty} \lambda_k e^{2\pi i k(s-t)}$$

and the following approximations can be used;

$$(\varphi_k)_j \sim \frac{1}{\sqrt{n}} e^{2\pi i k(j/n)}$$

$$\mu_k \sim n\lambda_k = n \int_0^1 q(t) e^{2\pi i k t} dt$$

and

$$\hat{d}_k = \sum_{j=1}^{n} (\varphi_k)_j d_j \sim \frac{1}{\sqrt{n}} \sum_{j=1}^{n} d_j e^{2\pi i k(j/n)} ,$$

with the fast Fourier transform being applied to evaluate μ_k and $\hat{d}_k$.

For this periodic situation, it is not difficult to show that regularization in the form above with $W = L^2$ is equivalent to transforming the given data into the frequency domain and then applying the low pass filter $1/(1+\alpha/\lambda_k)$. This equivalence can be exploited to obtain a different choice for α, as was done by Anderssen and Bloomfield [3], [4].

ACKNOWLEDGEMENT

I wish to thank Bob Anderssen, Frank de Hoog and Graeme Chandler for the dry run and subsequent comments, which have greatly improved the presentation of this paper.

REFERENCES

[1] R.A. Adams, Sobolev Spaces, Academic Press, 1975.

[2] R.S. Anderssen, On the use of linear functionals for Abel-type integral equations in applications, this book.

[3] R.S. Anderssen and P. Bloomfield, A time series approach to numerical differentiation, *Technometrics* 16 (1974), 69-75.

[4] R.S. Anderssen and P. Bloomfield, Numerical differentiation procedures for non-exact data, *Numer. Math.* 22 (1974), 157-182.

[5] P.M. Anselone and P.J. Laurent, A general method for the construction of interpolating or smoothing spline-functions, *Numer. Math.* 12 (1968), 66-82.

[6] N. Aronszajn, Theory of reproducing kernels, *Trans. Amer. Math. Soc.* 68 (1950), 337-404.

[7] C.T.H. Baker, The Numerical Treatment of Integral Equations, Oxford University Press, 1977.

[8] A.B. Bakushinskii, A numerical method for solving Fredholm equations of the first kind, *USSR Comput. Maths. Math. Phys.* 5 (1965), 226-233.

[9] C. de Boor and R.E. Lynch, On splines and their minimum properties, *J. Math. Mech.* 15 (1966), 953-969.

[10] S. Goldberg, Unbounded Linear Operators with Applications, McGraw-Hill, New York, 1966.

[11] J.W. Hilgers, On the equivalence of regularization and certain reproducing kernel Hilbert space approaches for solving first kind problems, *SIAM J. Numer. Anal.* 13 (1976), 172-184.

[12] J.W. Hilgers, Non-iterative methods for solving operator equations of the first kind, *MRC Technical Summary Report* #1413 (1974).

[13] F.R. de Hoog, Review of Fredholm equations of the first kind, this book.

[14] P. Linz, A general theory for the approximate solution of operator equations of the second kind, *SIAM J. Numer. Anal.* 14 (1977), 543-554.

[15] J. Locker and P.M. Prenter, Regularization with differential operators. I. General theory, *SIAM J. Math. Anal.*, to appear.

[16] J. Locker and P.M. Prenter, Regularization with differential operators. II. Weak least squares finite element solutions to first kind integral equations, *SIAM J. Numer. Anal.*, to appear.

[17] M.A. Lukas, in preparation.

[18] G. Ribiere, Regularization d'operateurs, *R.I.R.O.* 11 (1967), 57-79.

[19] I.J. Schoenberg, Spline functions and the problem of graduation, *Proc. Nat. Acad. Sci. U.S.A.* 52 (1964), 947-950.

[20] F. Smithies, Integral Equations, Cambridge University Press, London, 1958.

[21] A.N. Tikhonov, Solution of incorrectly formulated problems and the regularization method, *Soviet Math. Dokl.* 4 (1963), 1035-1038.

[22] A.N. Tikhonov, Regularization of incorrectly posed problems, *Soviet Math. Dokl.* 4 (1963), 1624-1627.

[23] A.N. Tikhonov and V.B. Glasko, An approximate solution of Fredholm integral equations of the first kind, *USSR Comput. Maths. Math. Phys.* 4 (1964), 564-571.

[24] G. Wahba, On the approximate solution of Fredholm integral equations of the first kind, *MRC Technical Summary Report* #990 (1969).

[25] G. Wahba, Practical approximate solutions to linear operator equations when the data are noisy, *SIAM J. Numer. Anal.* 14 (1977), 651-667.

A SURVEY OF METHODS FOR THE SOLUTION OF VOLTERRA INTEGRAL EQUATIONS OF THE FIRST KIND

Peter Linz
Mathematics Department, University of California, Davis, California

1. INTRODUCTION

We consider here the numerical solution of the linear Volterra equation of the first kind

1.1 $$\int_0^s k(s,t)f(t)dt = g(s) \ , \quad 0 \leqslant s \leqslant S \ .$$

We will restrict our discussion to cases for which the following assumptions hold

(a) $k(s,s) \neq 0$ for all $0 \leqslant s \leqslant S$,

(b) $g(0) = 0$,

(c) the functions $k(s,t)$ and $g(t)$ are bounded and sufficiently smooth (i.e., possess whatever derivatives are required in the subsequent analysis).

From an analytical viewpoint the most useful observation about (1.1) is that it can be recast as a Volterra equation of the second kind. For example, if we differentiate (1.1) we have

1.2 $$k(s,s)f(s) + \int_0^s k^{(1,0)}(s,t)f(t)dt = g'(s) \ ,$$

where we use the notation

$$k^{(i,j)}(s,t) = \left. \frac{\partial^{i+j}\, k(\xi,\zeta)}{\partial\xi^i\, \partial\zeta^j} \right|_{\xi=s,\ \zeta=t} .$$

Because of the assumption (a), equation (1.2) is of the second kind. Alternatively, if we introduce

1.3 $$\phi(s) = \int_0^s f(t)dt$$

and integrate (1.1) by parts, we obtain another equation of the second kind

1.4 $$k(s,s)\phi(s) - \int_0^s k^{(0,1)}(s,t)\phi(t)dt = g(s) .$$

The formal reduction of (1.1) to an equation of the second kind allows us to use the extensive body of results on equations of the second kind (see, for example, Kowalewski [16]) to study equations of the first kind. For numerical purposes this reduction, if possible, can also be quite useful, since there exist now many efficient methods for solving Volterra equations of the second kind. However, the reduction is not always possible. Often equations of this type occur in experimental situations (e.g., systems identification) and $k(s,t)$ and $g(s)$ may be known in tabular form only. In such cases it becomes of interest to study "direct" numerical methods, that is, methods which are based on (1.1) rather than on its reduced form.

There has been a considerable amount of work done on direct methods in the last ten years. It is the purpose of this paper to review the highlights of this development.

2. Some Simple Low-Order Methods

The construction of numerical methods for (1.1) is intuitively easy; we simply replace the integral by a numerical quadrature, then satisfy the equation at selected points in $0 \leq s \leq S$. Thus, we obtain a system of algebraic equations yielding an approximate solution. There are, of course, many possible ways in which this can be done, but as we will see, many of the obvious approaches do not work.

The Midpoint Method

We divide the interval $0 \leq s \leq S$ into N equal parts of width h. We will use the notation $t_\nu = \nu h$; F_ν will denote the computed approximation to $f(t_\nu)$. We now employ the composite mid-point rule (i.e., the integral of a function $\phi(t)$ between t_i and t_{i+1} is approximated by $h\phi(t_{i+\frac{1}{2}})$), and satisfy the resulting equation at $t_1, t_2, \ldots, t_N$. This yields the scheme

2.1 $$h\sum_{i=1}^{n} k(t_n, t_{i-\frac{1}{2}})\, F_{i-\frac{1}{2}} = g(t_n) , \quad n = 1,2,\ldots$$

which can be solved for consecutive values of n by

2.2 $$F_{n-\frac{1}{2}} = \frac{g(t_n)}{h\,k(t_n,t_{n-\frac{1}{2}})} - \sum_{i=1}^{n-1} \frac{k(t_n,t_{i-\frac{1}{2}})}{k(t_n,t_{n-\frac{1}{2}})} F_{i-\frac{1}{2}} , \quad n = 1,2,\ldots$$

From assumption (a) it follows that, for sufficiently small h, $k(t_n,t_{n-\frac{1}{2}}) \neq 0$ so that (2.2) always gives the solution of (2.1).

The effectiveness of the midpoint method is characterized by the following result.

THEOREM 1 *Assume that conditions (a), (b), (c) are satisfied, and in particular that* $k^{(0,3)}$, $k^{(1,1)}$ *and* $g^{(4)}$ *exist and are continuous in* $0 \leqslant t \leqslant s \leqslant S$. *Then the solution computed by (2.2) satisfies*

2.3 $$F_{n-\frac{1}{2}} - f(t_{n-\frac{1}{2}}) = \frac{h^2}{24} e(t_{n-\frac{1}{2}}) + O(h^3) ,$$

where $e(t)$ *is the solution of*

2.4 $$\int_0^s k(s,t)e(t)dt = \frac{\partial}{\partial t} [k(s,t)f(t)] \Big|_{t=0}^{t=s} .$$

Proof : See [17] or [18].

We see from this not only that the midpoint method has a second-order convergence, but also that the dominant part of the error is smooth. This plays an important part in the extrapolation techniques to be discussed below.

The Trapezoidal Method

Just as obviously we might use the trapezoidal method for the integration. In this case we obtain the scheme

2.5 $$\frac{h}{2} k(t_n,0)\ F_0 + h \sum_{i=1}^{n-1} k(t_n,t_i)\ F_i + \frac{h}{2} k(t_n,t_n)\ F_n = g(t_n) , \quad n = 1,2,\ldots$$

or

2.6 $$F_n = \frac{2}{h\,k(t_n,t_n)} \left\{ g(t_n) - \frac{h}{2} k(t_n,0)\ F_0 - h \sum_{i=1}^{n-1} k(t_n,t_i) F_i \right\} .$$

This procedure requires a starting value F_0 ; from (1.2) we see that we can use

2.7 $$F_0 = f(0) = \frac{g'(0)}{k(0,0)} .$$

THEOREM 2 *Under the assumptions of Theorem 1, the solution computed by (2.6) with* F_0 *given by (2.7), satisfies*

$$\max_{1\leq n\leq N} |F_n - f(t_n)| = O(h^2) .$$

Proof: See [15] and [17].

While this result guarantees second-order convergence for the trapezoidal method, it is much weaker than Theorem 1, in that it does not tell us whether the error behaves smoothly. In fact, a detailed analysis shows that the approximations produced by the trapezoidal method contain oscillations of the same order of magnitude as the error. The situation and a comparison with the midpoint method is depicted in Figure 1.

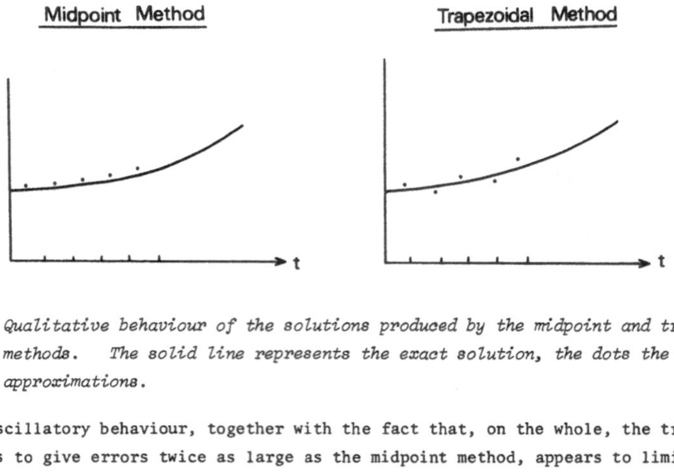

Figure 1. *Qualitative behaviour of the solutions produced by the midpoint and trapezoidal methods. The solid line represents the exact solution, the dots the computed approximations.*

This oscillatory behaviour, together with the fact that, on the whole, the trapezoidal method tends to give errors twice as large as the midpoint method, appears to limit the usefulness of the trapezoidal method.

The trapezoidal method appears to have been first suggested by Jones [14]; later it was completely analyzed by Kobayasi [15] and Linz [17].

3. The Construction of Higher Order Methods

Nonconvergence of Higher Order Quadrature Rules

When we attempt to construct higher order methods by using more accurate quadrature rules, such as the Gregory methods, or the Newton-Cotes rules (e.g., Simpson method and the three-eights rule), we find that the resulting schemes are unacceptable. They are all unstable, producing wildly oscillating results that become worse as we reduce h. This instability was first observed in [17], where some of the more obvious methods were investigated. Later results by Gladwin and Jeltsch [10] showed that this problem is general in that all of the standard rules (i.e., Gregory and Newton-Cotes) on equidistant points are unstable. Therefore, it is useless to look in this direction for more accurate algorithms.

Extrapolation from the Midpoint Method

The error estimate (2.3) provides the justification for the use of certain extrapolation procedures. For example, we can use *Richardson's extrapolation.* We first compute the solution via (2.2) using a stepsize h, then repeat the computation using a stepsize $3h$ (the factor of 3 is chosen so that the two sets of solutions are approximations at some common points). If for some fixed, s, we let $Y(h)$ and $Y(3h)$ denote the two computed solutions, we see from (2.3) that

$$Y(h) = f(s) + \frac{h^2}{24} e(s) + 0(h^3)$$

$$Y(3h) = f(s) + \frac{9h^2}{24} e(s) + 0(h^3) .$$

If we now compute the extrapolated approximation

3.1 $$Y^E = Y(h) + \frac{1}{8} [Y(h) - Y(3h)] ,$$

then

3.2 $$|Y^E - f(s)| = 0(h^3) .$$

Alternatively, we could use the method of *deferred corrections.* We first solve for F_n, then use this approximation in place of $f(t)$ in (2.4) and evaluate the right-hand side using finite differences. We solve (2.4) approximately for $e(t)$, using again (2.2), then finally use this to correct F_n. On the whole, however, this procedure seems more cumbersome than Richardson's extrapolation.

The trapezoidal method, on the other hand, is not suitable for extrapolation because of the oscillatory nature of the error.

Piecewise Polynomial Approximations and Block-by-Block Methods

The first general scheme for constructing methods of arbitrarily high order, the so-called block-by-block method (sometimes called Implicit Runge-Kutta method), was suggested and completely analyzed by de Hoog and Weiss (see [3], [4]). Here, instead of computing one value of F_n at a time, we find a block of several in each step. The idea is intuitively simple, but its description tends to become notationally cumbersome. We therefore illustrate the approach with a specific example.

Let us divide each interval $[t_n, t_{n+1}]$ into three parts using the points $t_{n1} = t_n + \frac{h}{3}$

and $t_{n2} = t_n + \frac{2h}{3}$. The approximate solution will be computed at the points t_{n1}, t_{n2} and t_{n+1}, for which we will use the symbols F_{n1}, F_{n2} and F_{n+1}, respectively. To obtain these unknown values we replace the integral by a numerical quadrature using these points and satisfy the equation at t_{n1}, t_{n2} and t_{n+1}. For example, we may use the quadrature formula

3.3 $$\int_{t_n}^{t_{n+1}} \phi(t)dt \simeq \frac{h}{4}\left\{3\phi(t_n + \frac{h}{3}) + \phi(t_{n+1})\right\}$$

(where $\phi(t)$ stands for an arbitrary integrand). When scaled to the smaller intervals these become

3.4 $$\int_{t_n}^{t_n + \frac{2h}{3}} \phi(t)dt \simeq \frac{h}{6}\left\{3\phi(t_n + \frac{2h}{9}) + \phi(t_n + \frac{2h}{3})\right\},$$

3.5 $$\int_{t_n}^{t_n + \frac{h}{3}} \phi(t)dt \simeq \frac{h}{12}\left\{3\phi(t_n + \frac{h}{9}) + \phi(t_n + \frac{h}{3})\right\}.$$

We want to apply these formulas to (1.1) so that $\phi(t)$ will be identified with $k(s,t)f(t)$. Since we will have approximations to $f(t)$ only at t_{n1}, t_{n2} and t_{n+1} everything must be expressed in terms of these values. The difficulty is easily overcome by using quadratic interpolation for $f(t)$.

3.6 $$f(t_n + \frac{h}{9}) \simeq \frac{1}{9}\left\{20f(t_n + \frac{h}{3}) - 16f(t_n + \frac{2h}{3}) + 5f(t_n + h)\right\},$$

3.7 $$f(t_n + \frac{2h}{9}) \simeq \frac{1}{9}\left\{14f(t_n + \frac{h}{3}) - 7f(t_n + \frac{2h}{3}) + 2f(t_n + h)\right\}.$$

Using these approximations in the original equation and satisfying the result at t_{n1}, t_{n2} and t_{n+1} we are led to the system

3.8 $$\frac{k(t_{n1}, t_n + \frac{h}{9})}{36}\left\{20F_{n1} - 16F_{n2} + 5F_{n+1}\right\} + \frac{k(t_{n1}, t_{n1})}{12} F_{n1} = \frac{g(t_{n1})}{h} - \sum_{i=1}^{n-1}\left\{\frac{3}{4}k(t_{n1}, t_{i1})F_{i1} + \frac{1}{4}k(t_n, t_{i+1})F_{i+1}\right\},$$

3.9 $$\frac{k(t_{n2}, t_n + \frac{2h}{9})}{18}\left\{14F_{n1} - 7F_{n2} + 2F_{n+1}\right\} + \frac{k(t_{n2}, t_{n2})}{6} F_{n2} = \frac{g(t_{n2})}{h} - \sum_{i=0}^{n-1}\left\{\frac{3}{4}k(t_{n2}, t_{i1})F_{i1} + \frac{1}{4}k(t_{n2}, t_{i+1})F_{i+1}\right\},$$

3.10 $$\frac{3}{4}k(t_{n+1}t_{n1})F_{n1} + \frac{1}{4}k(t_{n+1}, t_{n+1})F_{n+1} = \frac{g(t_{n+1})}{h} - \sum_{i=1}^{n-1}\left\{\frac{3}{4}k(t_{n+1}, t_{i1})F_{i1} + \frac{1}{4}k(t_{n+1}, t_{i+1})F_{i+1}\right\}.$$

These three simultaneous equations are solved for $n = 1,2,\ldots$ to yield a block of three unknowns F_{n1}, F_{n2} and F_{n+1} at a time.

It can be shown that the method described by (3.8) - (3.10) has order of convergence three. The scheme can obviously be extended and the analysis of de Hoog and Weiss shows that such schemes are convergent with the accuracy depending on the number and the spacing of the intermediate points.

Closely related to the block-by-block methods are the schemes devised by piecewise polynomial approximation of the unknown. As above, we will assume that we have subdivided the interval $[t_n, t_{n+1}]$ by the points $t_n \leq t_{n1} < t_{n2} < \ldots < t_{np} \leq t_{n+1}$. In each of the intervals $[t_n, t_{n+1}]$ we will approximate the solution by a polynomial $\phi_n(t)$ of degree p-1. Let us rewrite (1.1) as

3.11 $$\int_{t_n}^{s} k(s,t)f(t)dt = g(s) - \int_0^{t_n} k(s,t)f(t)dt .$$

In general, no piecewise polynomial approximation can satisfy this equation exactly, which leaves a number of ways of choosing $\phi_n(t)$ so as to obtain approximate equality. One of the most convenient ways is to satisfy (3.11) at the points t_{ni} (collocation), which leads to

$$\int_{t_n}^{t_{ni}} k(t_{ni},t)\phi_n(t)dt = g(t_{ni}) - \sum_{\ell=0}^{n-1} \int_{t_\ell}^{t_{\ell+1}} k(t_{ni},t)\phi_\ell(t)dt ,$$

3.12 $$i = 1,2,\ldots p, n = 0,1,2,\ldots .$$

The computational procedure involved here becomes clearer if we explicitly represent $\phi_n(t)$ in terms of powers of t. Let us write

3.13 $$\phi_n(t) = \alpha_{n1} + \alpha_{n2}\, t + \alpha_{n3}\, t^2 + \ldots + \alpha_{np}\, t^{p-1} .$$

Substituting this into (3.12), we obtain the linear system

3.14 $$M_n \underset{\sim}{\alpha}_n = \underset{\sim}{g}_n - \sum_{\ell=0}^{n-1} N_\ell \underset{\sim}{\alpha}_\ell$$

where

$$\underset{\sim}{\alpha}_n = \begin{pmatrix} \alpha_{n1} \\ \alpha_{n2} \\ \vdots \\ \alpha_{np} \end{pmatrix} , \quad \underset{\sim}{g}_n = \begin{pmatrix} g(t_{n1}) \\ g(t_{n2}) \\ \vdots \\ g(t_{np}) \end{pmatrix} ,$$

and M_n and N_ℓ are pxp matrices with elements

3.15 $$M_n(i,j) = \int_{t_n}^{t_{ni}} k(t_{ni},t)t^{j-1}dt ,$$

3.16 $$N_\ell(i,j) = \int_{t_\ell}^{t_{\ell+1}} k(t_{ni},t)t^{j-1}dt .$$

The use of piecewise polynomial approximations has been studied in detail by Brunner (see [1], [2]), who showed that with properly chosen collocation points such methods generally yield convergent, higher order algorithms. It was also pointed out by Brunner that the block-by-block methods can be obtained by this approach if we use particular numerical quadratures to approximate the moment integrals (3.15) and (3.16).

It should be noted that the construction used here imposes no continuity requirement on the approximate solution, and generally there will be small jump discontinuities at the points t_n. However, this seems to be of little consequence, as the magnitude of these jumps is of the same order as the error itself. In fact, in solving Volterra equations of the first kind, it is best to impose as few continuity requirements on the approximate solution as possible. One can construct methods which produce continuous solutions but these, although convergent, tend to show small oscillations such as noted in connection with the trapezoidal method. If we try to impose more stringent smoothness conditions, such as approximating $f(t)$ by a cubic spline, we obtain unstable algorithms unless special steps are taken to ensure stability. (For a discussion of these points see [6], [7], [13]).

The Use of Nonstandard Quadratures

While all the familiar quadrature rules yield unstable algorithms, it is possible to use specially designed formulae to produce usable methods. Basically, these special methods sacrifice some accuracy in order to achieve stability. The general idea is best explained with a specific example.

Suppose that we have decided to use the three-eights rule

$$\int_0^{3h} \phi(t)dt \cong h\{\tfrac{3}{8}\,\phi(0) + \tfrac{9}{8}\,\phi(h) + \tfrac{9}{8}\,\phi(2h) + \tfrac{3}{8}\,\phi(3h)\}$$

as the basic integration rule. When applied in composite form it can be used as it stands only when $n = 3m+1$. For $n = 3m+2$ and $n = 3m+3$ some adjustment has to be made, for example by using a different formula near the upper end of the interval. If we leave these endpoint formulae unspecified for the moment, then the weights will have the pattern shown in Figure 2.

$n = 3m+1$	$\frac{3}{8}$	$\frac{9}{8}$	$\frac{9}{8}$	$\frac{6}{8}$	$\frac{9}{8}$	$\frac{9}{8}$	. . . .	$\frac{6}{8}$	$\frac{9}{8}$	$\frac{9}{8}$	$\frac{3}{8}$			
$n = 3m+2$	$\frac{3}{8}$	$\frac{9}{8}$	$\frac{9}{8}$	$\frac{6}{8}$	$\frac{9}{8}$	$\frac{9}{8}$	. . . .	$\frac{3}{8}+\alpha_1$	α_2	α_3	α_4	α_5		
$n = 3m+3$	$\frac{3}{8}$	$\frac{9}{8}$	$\frac{9}{8}$	$\frac{6}{8}$	$\frac{9}{8}$	$\frac{9}{8}$	. . . .	$\frac{3}{8}+\beta_1$	β_2	β_3	β_4	β_5	β_6	

Figure 2. *Pattern of weights for three-eights rule with end adjustment.*

We now must determine $\alpha_1 \ldots, \alpha_5$ and $\beta_1, \ldots, \beta_6$. If we were to select these so as to obtain maximum accuracy, we could make the rule involving the α's exact for polynomials of degree 4 and hence locally of error $O(h^6)$, while the rule involving the β's could be made exact for polynomials of degree 5 and hence would have a local error of $O(h^7)$. However, if we did this, the resulting procedure would be unstable. We therefore relax the accuracy requirements, asking only that both rules have local error $O(h^5)$, which is sufficient and consistent with the three-eights rule which also has a local $O(h^5)$ error. These requirements impose four conditions on the α's and four conditions on the β's, leaving three undetermined parameters. These three parameters are then chosen so as to make the method stable. It is not easy to see that this last step can be successfully carried out, and unfortunately a demonstration would require carrying out a lengthy error analysis. Suffice it therefore to say that the approach does work in the example given as well as in many other cases. For details we must refer the reader to Holyhead, McKee and Taylor [11] and Holyhead and McKee [12] where a complete discussion can be found. Related ideas are discussed in Gladwin ([8],[9]) and Taylor [20].

Discretization of Differentiated Forms

Finally, it should be mentioned that the equivalent equations of the second kind (1.2) and (1.4) may be useful for numerical purposes even if the derivatives cannot be found explicitly. All we need to do is to use finite difference approximations for these derivatives and then employ a standard method for equations of the second kind. The procedure is conceptually straightforward and a rigorous justification is not difficult. However, whether such an approach is competitive with the direct methods has not been completely investigated.

4. The Effect of Computational and Experimental Errors

Although we have shown how one can construct highly accurate methods for the solution of Volterra equations of the first kind, there is one point we have so far ignored: the effect of computational and experimental errors on the numerical scheme. There is of course always some round-off error in any lengthy computation; furthermore, if (1.1) arises from an experimental situation then the observational errors will affect the solution. Therefore, we must consider the effect of small perturbations on the computed solution.

One reason why this point needs to be of particular concern is that (1.1) is not well-posed! We can see this from (1.2), which shows that f depends on g'. Thus a small change in g can cause an arbitrarily large change in f. Some authors (e.g.[5],[19]) have recognized this problem and have proposed standard regularization methods for the solution of (1.1). However, the situation for Volterra integral equations is not nearly as bad as the one encountered with Fredholm equations of the first kind, and it is not clear that one needs to adopt the extreme remedy of complete regularization.

To investigate this problem further, let us reconsider the midpoint method (2.1), perturbing the right-hand side by an amount η_n. The corresponding perturbed solution $\hat{F}_n$ will then be given by

$$4.1 \qquad h \sum_{i=1}^{n} k(t_n, t_{i-\frac{1}{2}})\, \hat{F}_{i-\frac{1}{2}} = g(t_n) + \eta_n \, .$$

The relation between $F_{n-\frac{1}{2}}$ and $\hat{F}_{n-\frac{1}{2}}$ is easily established.

THEOREM 3 *Let* $F_{n-\frac{1}{2}}$ *and* $\hat{F}_{n-\frac{1}{2}}$ *denote the respective solutions of (2.1) and (4.1) . If*

$$\max_{1 \leq n \leq N} |\, \eta_n \,| \leq \eta$$

and if the conditions of Theorem 1 are satisfied then, for $h < h_0$

$$4.2 \qquad |\, F_{n-\frac{1}{2}} - \hat{F}_{n-\frac{1}{2}} \,| \leq \frac{2\eta}{hE}\, e^{(n-1)h\, C/E}$$

where $C = \max_{0 \leq t \leq s \leq S} |\, k^{(1,0)}(s,t) \,|$ and $E = \min_{\substack{0 \leq s \leq S \\ 0 \leq h \leq h_0}} |\, k(s, s-h/2) \,|$.

Proof: Let $\Delta_j = F_{j-\frac{1}{2}} - \hat{F}_{j-\frac{1}{2}}$. Then, subtracting (1.28) from (1.5) we have

$$h \sum_{j=1}^{n} k(t_n, t_{j-\frac{1}{2}})\, \Delta_j = \eta_n .$$

This equation is in a form which arises in the analysis of the midpoint method itself, so that (4.2) can be deduced by applying the technique used in [18].

The bound (4.2) , although possibly pessimistic for some cases, is attained for the simple case $k(s,t) = 1, \eta_n = (-1)^n \eta$. Thus, in the worst case the error is magnified by a factor proportional to $1/h$. This error magnification is inherent in the problem, and is not due to any defect in the midpoint method. Any other method we might try leads to similar (or perhaps even worse) results. For example, if we try to solve the equivalent second kind equations (1.2) or (1.4) then the numerical differentiation involved will again magnify any error in g by a factor proportional to $1/h$. Thus, whatever method we use there will be error magnification; fortunately it is of a limited extent and generally manageable. From (4.2) we can draw the following conclusions:

(a) Unless very high accuracy is required, round-off error is not likely to present a serious problem. For example, on a typical computer, the error due to round-off might be of the order 10^{-8} or 10^{-9}. Practical limitations make it unlikely that we would use an h much smaller than 10^{-3}. Thus, we should be able to obtain an accuracy of 10^{-5} to 10^{-6}, which is sufficient in most practical problems.

(b) The inaccuracy due to experimental error may be more serious. If this error is about 1% of the true value, then for $h = 10^{-2}$ the effect on the solution may be of the same order of magnitude as the solution itself, completely invalidating the results.

Since this last point can cause serious difficulties, considerable care has to be taken when observational errors are present. Elementary precautions, such as smoothing of all experimental data, should undoubtedly be used. However, the general question of what methods are the most appropriate for the solution of Volterra integral equations in the presence of experimental errors has received little attention and seems to be an open problem.

References

[1] Brunner, H. Discretization of Volterra equations of the first kind, *Math. Comp.*, 31, 708-716, 1977.

[2] Brunner, H. Discretization of Volterra equations of the first kind (II), *Num. Math*, 30, 117-136, 1978.

[3] De Hoog, F. and Weiss, R. On the solution of Volterra integral equations of the first kind, *Num. Math.*, 21, 22-32, 1973.

[4] De Hoog, F. and Weiss, R. High order methods for Volterra integral equations of the first kind, *SIAM. J. Numer. Anal.*, 10, 647-664, 1973.

[5] Douglas, J. Jr. Mathematical programming and integral equations, in *Symposium on the Numerical Treatment of Ordinary Differential Equations, Integral and Integrodifferential Equations*, Proc. Rome Symposium PICC, Birkhäuser, Basel, 269-274, 1960.

[6] El Tom, M.E.A. On spline function approximations to the solution of Volterra integral equations of the first kind, *BIT*, 14, 288-297, 1974.

[7] El Tom, M.E.A. Application of spline functions to systems of Volterra integral equations of the first and second kinds, *J. Inst. Math. Applic.*, 17, 295-310, 1976.

[8] Gladwin, C.J. Methods of higher order for the numerical solution of first kind Volterra integral equations, Proc. Second Manitoba Conference on Numerical Mathematics, 179-193, 1972.

[9] Gladwin, C.J. Numerical solution of Volterra equations of the first kind, Ph.D. Thesis, Dalhousie University, 1975.

[10] Gladwin, C.J. and Jeltsch, R. Stability of quadrature rule methods for first kind Volterra integral equations, *BIT*, 14, 144-151, 1974.

[11] Holyhead, P.A.W., McKee, S. and Taylor, P.J. Multistep methods for solving linear Volterra integral equations of the first kind, *SIAM. J.Numer.Anal.*, 12, 698-711, 1975.

[12] Holyhead, P.A.W. and McKee, S. Stability and convergence of multistep methods for linear Volterra integral equations of the first kind, *SIAM. J. Numer. Anal.*, 13, 269-292, 1976.

[13] Hung, H.S. The numerical solution of differential and integral equations by spline functions, Technical Summary Report 1053, Mathematics Research Center, University of Wisconsin, 1970.

[14] Jones, J.G. On the numerical solution of convolution integral equations and systems of such equations, *Math. Comp.*, 15, 131-142, 1961.

[15] Kobayasi, M. On the numerical solution of the Volterra integral equation of the first kind by the trapezoidal rule, *Rept. Stat. Appl. Res. Japan*, 14, 1-14, 1967.

[16] Kowalewski, G. *Integralgleichungen*, de Gruyter, Berlin 1930.

[17] Linz, P. The numerical solution of Volterra integral equations by finite difference methods, Technical Summary Report 825, Mathematics Research Center, University of Wisconsin, 1967.

[18] Linz, P. Numerical methods for Volterra integral equations of the first kind, *Comput. J.*, 12, 393-397, 1969.

[19] Schmaedeke, W.W. Approximate solutions of Volterra integral equations of the first kind, *J. Math. Anal. Appl.*, 23, 604-613, 1968.

[20] Taylor, P.J. The solution of Volterra integral equations of the first kind using inverted differentiation formulae, *BIT*, 16, 416-425, 1976.

ON THE USE OF LINEAR FUNCTIONALS FOR ABEL—TYPE INTEGRAL EQUATIONS IN APPLICATIONS

R.S. Anderssen
Computing Research Group/Pure Mathematics, SGS, Australian National University, and Division of Mathematics and Statistics, CSIRO, Canberra, A.C.T.

Abstract

A commonly occurring class of Volterra integral equations which arise regularly in applications are the separable first kind Abel-type integral equations such as

$$s(y) = \int_y^a k_1(y)k_2(x)(x^2-y^2)^{-\mu}u(x)dx\ , \quad 0 < \mu < 1\ , \quad 0 \le y \le x \le a \le \infty\ . \qquad (*)$$

However, it is often linear functionals (e.g. moments) defined on the unknown $u(x)$ *, namely*

$$m_\theta(u) = \int_0^a \theta(t)u(t)dt\ ,$$

which are used for inference purposes not $u(t)$ *itself. Numerically, the direct evaluation of such functionals poses major difficulties since it is first necessary to obtain an accurate numerical approximation to* $u(t)$ *when* $s(y)$ *is only available as a series of discrete observations*

$$\{d_i\} = \{d_i = s(x_i) + \varepsilon_i\ ,\ \ i = 1,2,\ldots,n;\ \varepsilon_i \ \text{ discrete random errors}\}\ .$$

If such functionals are evaluated as functionals defined on the given data $\{d_i\}$ *, namely*

$$m_\phi(s) = \int_0^a \phi(\theta;t)s(t)dt\ ,$$

then difficulties associated with the numerical solution of (*) are avoided. The utility of this strategy is examined in some detail.

1. Introduction

Volterra integral equations

$$s(y) = \lambda u(y) + \int_y^a K(y,x)u(x)dx \ , \quad 0 \le y \le x \le a \le \infty \ , \tag{1.1}$$

for which the kernel $K(y,x)$ takes the form

$$K(y,x) = k(y,x)/(x^p - y^p)^\mu \ , \quad 0 < \mu < 1 \ , \quad p > 0 \ , \tag{1.2}$$

where $k(y,x)$ is continuous on $0 \le y \le x \le a$ with $k(y,y) \neq 0$, are often referred to as *Abel-type integral equations*. They and mathematically equivalent forms occur regularly in such diverse areas as biology, metallurgy, microscopy, astrophysics, solar physics and seismology.

A survey of many of these applications can be found in Anderssen [2]. As shown there, a majority can be classified as first kind Abel-type equations with separable kernels; viz.,

$$\int_y^a k_1(y)k_2(x)(x^2-y^2)^{-\frac{1}{2}}u(x)dx = s(y) \ . \tag{1.3}$$

Since the separable parts of the kernel can be absorbed into the signal $s(y)$ and the solution $u(x)$ to yield

$$s(y) = \int_y^a (x^2-y^2)^{-\frac{1}{2}}u(x)dx \ , \tag{1.4}$$

we limit attention to an examination of (1.4). Numerically, such equations are of considerable interest, not only because they arise regularly in applications, but also because they are weakly improperly posed and therefore require the use of more sophisticated numerical schemes than might at first sight seem unnecessary. This point assumes added importance because, in most applications, the data is not available as a function but are observations made at discrete points, viz.

$$\{d_i\} = \{d_i = s(x_i) + \varepsilon_i \ , \quad x_1 < x_2 < \ldots < x_n \ ; \ \varepsilon_i \ \text{random errors}\} \ . \tag{1.5}$$

In fact, the numerical problem is not simply to solve (1.4) for given $s(t)$; but, to solve (1.4) with respect to given discrete observations of $s(t)$ (namely, $\{d_i\}$). When in fact the data $\{d_i\}$ corresponds to observations of a cumulative distribution, for which $s(y)$

denotes the corresponding density distribution, either (1.4) has to be modified, or the data $\{d_i\}$ must be differentiated numerically to yield an estimate of $s(x_i)$, $i = 1,2,\ldots,n$, which we shall denote by $\{d_i\}$. In the sequel, only the latter strategy will be examined.

In many applications for which the Abel-type formalism (1.4) applies, it is often linear functionals defined on the unknown solution $u(x)$, viz.

$$m_\theta(u) = \int_0^a (x)u(x)dx\ , \quad 0 < a \le \infty\ , \tag{1.6}$$

which are used for inference purposes not $u(t)$ itself. Exemplification of this point will be discussed in §2, while numerical results for real and synthetic data will be given in §3.

Numerically, the direct evaluation of such functionals often poses major difficulties since it is first necessary to obtain an accurate numerical approximation to $u(t)$. If such functionals could be redefined as functionals on the signal $s(t)$ (the exact date), viz.

$$m_\phi(s) = \int_0^a \phi(x)s(x)dx\ , \quad 0 < a \le \infty\ , \tag{1.7}$$

then the discrete observational data $\{d_i\}$ could be used to evaluate them directly. As a consequence, difficulties associated with the numerical solution of the original Abel equation would be circumvented.

In the sequel, we shall refer to the direct evaluation of (1.6) as the *direct strategy* and the use of (1.7) as the *data functionals strategy*.

The utility of the latter strategy for first kind Abel-type equations (of a form which can be transformed to (1.4)) is examined in some detail in §3. The key property of (1.4) via which functionals of the form (1.6) can be replaced by corresponding functionals of the form (1.7) is the existence of the analytic inversion formulas (under the assumption that $s(x) \equiv 0$, $x \ge a$)

$$u(x) = -\frac{1}{\pi}\int_x^a s'(y)(y^2-x^2)^{-\frac{1}{2}}dy\ , \tag{1.8}$$

$$= -\frac{1}{\pi x}\frac{d}{dx}\int_x^a ys(y)(y^2-x^2)^{-\frac{1}{2}}dy\ , \tag{1.9}$$

where $s'(y) = ds(y)/dy$.

The extension of the data functionals strategy to linear functionals defined on the solution of general integral equations is discussed briefly in §4.

§2. Exemplification for the Use of Linear Functionals for Abel-Type Integral Equations

To illustrate, we consider the following problem: *for a given consignment of steel, determine whether it has the (impact) strength to do a given job.* As the result of extensive experimentation, metallurgists claim that (Hyam and Nutting [8])

(i) *Steel consists of* a conglomerate of ferrite grains (composed of ferrous crystals) and carbide (carbon-type) particles, which to a first approximation can be regarded as spheres.

(ii) *The strength of steel depends on its hardness.* If it is too hard, it is too brittle. Tempering is used to make it softer and thereby increase its strength.

(iii) *The hardness of tempered steels depends upon the ferrite grain sizes.* The general rule is "the larger the ferrite grains, the softer the steel".

(iv) *The growth of the ferrite grains is limited by the carbide particles.* The carbide particles form along grain boundaries, and therefore the size of the ferrite grains is heavily constrained by the size distribution and number density of the carbide particles (assuming the centers of the particles are distributed something like a Poisson process).

(v) *During tempering, the number density of carbide particles decreases but their sizes grow, so the ferrite grains grow.* As tempering progresses, the microstructure and mechanical properties continue to alter, due to the development of approximately spherical particles of carbide (cementite) which gradually grow in size and decrease in number.

The mentioned change in size and number of carbide particles can only occur by the solution of the smaller particles, diffusion of carbon through the ferrite, and the simultaneous growth of the larger particles. This indicates that, because the amount of ferrite, carbon, etc. in a given consignment of steel being tempered is fixed, there is *a trade-off between number density* ρ *and size distribution* $u(x)$ of carbide particles during tempering at the expense of hardness and hence at the gain of strength.

Thus, since a knowledge of the size distribution $u(x)$ is necessary for estimating the number density ρ , the problem is reduced to an examination of the properties of the size distribution $u(x)$. This completes the initial analysis of the application in which *the properties of the application which relate directly to the problem of interest are pinpointed.*

Next we derive a *mathematical formalism* which relates $u(x)$ to observable data. As a first approximation, we shall assume that the carbide particles are spherical. Since it is not possible to observe the carbide spheres and hence $u(x)$ directly, it is necessary to estimate it by some indirect procedure. It is because of this necessity to observe the thing one requires by some indirect procedure that improperly posed formulations often arise. The ways in which this can be done are not unique, and depend heavily on the sampling strategy used.

The one which we shall pursue here is to use the size distribution of circles $s(y)$ (i.e., the radius or diameter distribution) defined on random plane sections taken through the steel. The sampling strategy is: on random plane (polished) sections taken through a suitably chosen sample of the steel, measure the observed size distribution of circles. This will yield the data $\{d_i\}$.

The motivation for assuming first that the carbide particles are spherical (as well as agreeing approximately with observation) and for taking the size distribution of circles $s(x)$, as the means by which we estimate $u(x)$, is the existence of the following mathematical formalism (an Abel-type integral equation) which relates $u(x)$ and $s(y)$:

$$s(y) = \frac{y}{m}\int_y^a u(x) \;/\; (x^2-y^2)\, dx \ , \quad 0 \le y \le x \le a \le \infty \ , \tag{2.1}$$

where $[0,a]$ defines the support of $u(x)$ and

$$m = \int_0^a xu(x)dx = \frac{\pi}{2}\left(\int_0^a \frac{s(x)}{x}dx\right)^{-1} \tag{2.2}$$

denotes the average sphere radius.

Note 2.1. The actual procedure, for choosing a sample of steel (from the consignment), and then taking random plane sections through it, is non-trivial and involves a number of deep (probabilistic) questions. A discussion of some of the difficulties which arise along with proposals for using volume weighted and area weighted sampling can be found in Miles and Davy [15] and [16].

The sampling and measurement of circles will yield discrete observations of the cumulative distribution

$$S(y) = \int_0^y s(z)dz \ , \quad S(0) = 0 \ , \quad S(a) = 1 \ ,$$

namely

$$d_i = S(y_i) + \varepsilon_i \ , \quad 0 < y_1 < y_2 < \dots < y_n \ .$$

We assume that the measurement process is sufficiently accurate so that duplicate observations do not occur. Once n is sufficiently large (so that a sufficiently detailed amount of information about the structure of $S(y)$ has been accumulated), a viable computational strategy can be applied to the data $\{d_i\}$ to generate an approximation

$$\{\hat{u}_j\} = \{\hat{u}(x_j); \ \ x_j = (j-1)\ x \ , \ \ j = 1,2,\dots,N \ , \ \Delta x = a/(N-1) \ , \ N < n\}$$

to $u(x)$ on an even grid $\{x_j\}$. For example, apply a two-stage procedure along the lines proposed by Anderssen [1], implemented by Anderssen and Jakeman [4], and discussed in §4 of Anderssen [2].

The approximation $\{u_i\}$ is then used to derive inferences about the strength of the steel in the given consignment. But, here is the catch. Though the metallurgist argues that the strength of steel is governed by the size distribution $u(x)$ of the carbide particles, he doesn't necessarily use $u(x)$ directly {see, for example, Hilliard [6], p.1373, Underwood [17], p.149, Hyam and Nutting [8]} when making a decision about whether a given steel has the strength to do a given job. Instead, he makes inferences on the basis of the values of functionals defined on $u(x)$. The functionals of $u(x)$ commonly used by metallurgists for this purpose are (see, for example, Hyam and Nutting [8]) :

a. average spheres radius m of (2.2) (which we assumed above to be known);

b. average particle surface area

$$A_u = 4\pi \int_0^a x^2 u(x)\,dx \; ; \tag{2.3}$$

c. average particle volume

$$V_u = \frac{4\pi}{3} \int_0^a x^3 u(x)\,dx \; ; \tag{2.4}$$

d. number of particles N_V per unit volume (assuming their centers follow a Poisson distribution)

$$N_V = N_A \,/\, (2m) \; , \tag{2.5}$$

where N_A denotes the number of circles per unit area of planar section of the sample of steel on which the data $\{d_i\}$ was measured.

Thus, for inference purposes, estimates of the functionals m , A_u and V_u , and whence N_V , are required. Often, metallurgists estimate m , A_u and V_u by evaluating the integrals (2.2), (2.3) and (2.4) using the approximation $\{\hat{u}_i\}$ - see, for example, Hyam and Nutting [8], p.151, column 2. They miss the point that, if only estimates of m , A_u , V_u and N_V are required for inference purposes, then the necessity to explicitly estimate $u(x)$ can be circumvented.

The second inversion formula for (1.4), viz. (1.9), can be used to replace linear functionals defined on $u(x)$, namely

$$m_\theta(u) = \int_0^a \theta(x)u(x)\,dx \; ,$$

by linear functionals on $s(x)$, namely

$$m_\phi(u) = \int_0^a \phi(x)s(x)\,dx \; .$$

In fact, for a given $\theta(x)$, the corresponding $\phi(x)$ is given by

$$\phi(x) = \frac{2m}{\pi} \left\{ \int_0^a \frac{d\theta}{dy} (x^2-y^2)^{-\frac{1}{2}} dy \right\} , \tag{2.6}$$

if we assume that $\theta(x)$ is such that $\frac{d\theta}{dy}(x^2-y^2)^{-\frac{1}{2}}$ is integrable, $\theta(0) = 0$ and $\theta(y) \neq y$. When $\theta(y) = y$, it is necessary to employ the type of independent argument used by Watson [18], since the above formula (2.6) for $\phi(x)$ involves m . The existence of the general correspondence (2.6) between $\theta(x)$ and $\phi(x)$ appears to have been overlooked in practice, except for the special cases examined by Meisner [13], Watson [18] and Jakeman [9], eqn(4.3.1).

Thus, the problem of estimating the functionals (2.2), (2.3), (2.4) and (2.5) reduces to evaluating the functionals

$$m = \frac{\pi}{2} \left(\int_0^a \frac{s(x)}{x} dx \right)^{-1} \tag{2.7}$$

$$A_u = 16m \int_0^a xs(x)dx \tag{2.8}$$

$$V_u = 2\pi m \int_0^a x^2 s(x)dx \tag{2.9}$$

$$N_V = N_A \ / \ (2m) \tag{2.10}$$

which are properly posed formulations. Though standard quadrature rules, such as the mid-point and Simpson, could be applied to evaluate the above functionals using the data $\{d_i\}$, it is important to take other relevant factors into account. Not only is it necessary to choose a quadrature which performs well numerically; it is also necessary, because of the observational nature of the data $\{d_i\}$ {see eqn(1.5)}, to choose quadrature rules which yield estimators of the above functionals which have appropriate statistical properties. This problem has been examined in some detail by Anderssen and Jakeman [3] and Jakeman and Scheaffer [12]. They conclude that, especially for functionals involving integrals of the form

$$\int_z^a \frac{s(x)}{(x^2-z^2)^{\frac{1}{2}}} dx \ , \quad 0 \le z < a \ ,$$

product integration procedures should be used.

§3. The Utility of the Data Functionals Strategy

In order to examine the utility of the data functionals strategy, we compare it with the direct strategy for the two independent situations where the given data (i) arises in applications, and (ii) is synthetic. An example of real sample distribution data is given in Figure 1.

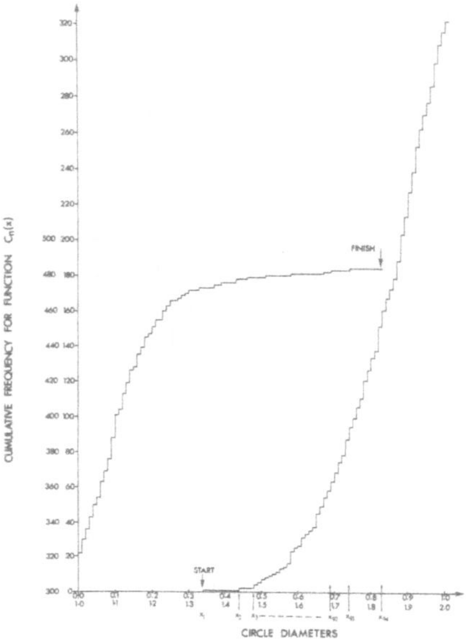

FIGURE 1.

Initially, we show that, when both strategies are applied to the same observational data to compute certain linear functionals defined on the solution of (2.1), the resulting estimates do not necessarily agree. Unfortunately, from the point of view of such comparisons, few authors publish sufficient details about their data to allow them to be used to check the results the authors claim to have obtained from them. One exception is the work of Meisner [13], in which both real and synthetic data is examined. For this reason, attention will be limited to an examination of these data.

Before comparisons can be made, it is necessary to construct algorithms for the evaluation of the data functionals

$$m_\phi(s(y)) = \int_0^a \phi(y)s(y)dy , \quad a < \infty , \tag{3.1}$$

with respect to the given data $\{d_i\}$ (cf. (1.5)). In fact, it is necessary to distinguish between the two situations where the data $\{d_i\}$ corresponds to discrete observations of either a density or a cumulative distribution. We shall refer to these two types of data as *discrete density data* and *discrete cumulative data*, respectively.

The basic strategy is to use product integration. For the evaluation of

$$m_\phi(s(y)) = \int_0^a \phi(y)s(y)dy \ , \quad a < \infty \ ,$$

with

$$d_i = s(y_i) + \varepsilon_i \ , \quad 0 \le y_0 < y_1 < \ldots < y_n \le a \ ,$$

where the ε_i denote random errors, the aim is to use the data $\{d_i\}$ to construct an approximation $s_n(y)$ to $s(y)$ so that $m_\phi(s_n(y))$ can be determined analytically. Details of the various algorithms compared and discussed in the sequel (viz. PMP#HT, PMP#T and PT) are given in Appendix A.

The data in Meisner [13] consists of two synthetic data sets (see Appendix B for details) and one observed diameter distribution as determined from the photomicrograph of a plane section through a sample of high-impact polystyrene. The former correspond to the grouped (discrete density) data listed in Table I and II in Meisner. They will be referred to as MEISNER#1 and MEISNER#2. The observed diameter distribution is given in two forms. In Table III, it is given as the observed sample cumulative (discrete cumulative data), while in Table IV it has been grouped to yield discrete density data (see Figure 2). They will be referred to as MEISNER#3 and MEISNER#4. The advantage of having the observed diameter distribution in these two forms is that it allows a comparison to be made of the effect of grouping sample cumulative data.

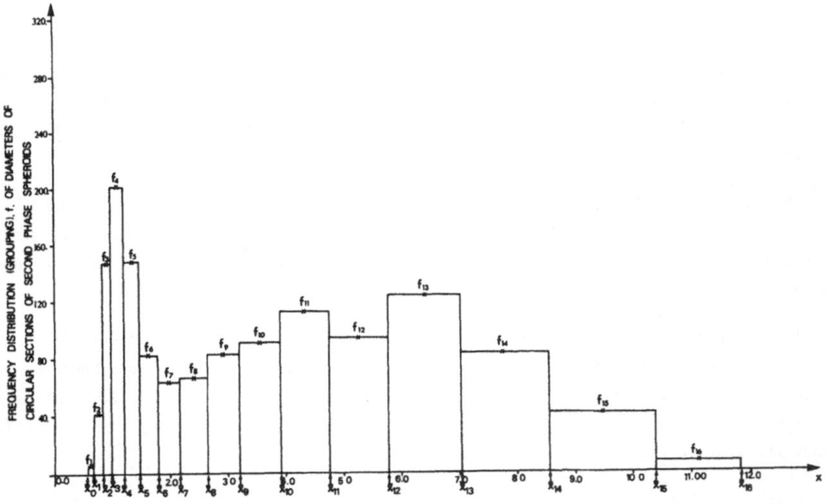

FIGURE 2

Meisner's direct method consists in applying a product integration method {similar to Wicksell's [19]} to solve (2.1) with respect to given discrete density data for $s(y)$ to yield a piecewise constant approximation to $u(y)$. With respect to this type of approximation, linear functionals with $\theta(x) = x^p$, $p > -1$, can be evaluated analytically. We shall denote this method by "M".

Meisner did in fact observe and utilize the result that the linear functionals (1.6) with $\theta(x) = x^p$, $p \geq 0$, namely

$$M_p = \int_0^a x^p u(x)dx\ , \qquad \bar{M}_p = \int_0^a (x-m)^p u(x)dx\ , \quad p \geq 0\ ,$$

could be evaluated using the data $\{d_i\}$ as a combination of functionals of the form (1.7), namel)

$$D_q = \int_0^a x^q s(x)dx\ , \quad q \geq -1\ ;$$

but he did not describe the algorithm used. In the comparisons given below, these results are included under the heading "MDF".

The algorithms PMP#HT, PMP#T and PT were applied to the discrete synthetic data MEISNER#1 and MEISNER#2 to compute the functionals given in Table 1. They are listed in Tables 2 and 3 (respectively). For comparison purposes, the available estimates of Meisner are also shown.

On the basis of these results, the following conclusions can be drawn :

1. Excluding kurtosis, a comparison of the M and MDF estimates with the "exact values" indicates that functionals evaluated on the data $\{d_i\}$ yield results which compare as well or better with the true than the corresponding functionals evaluated using the direct strategy.

2. The large errors occurring in both the M and MDF estimates of kurtosis would appear to imply the existence of gross errors in the algorithms used for its computation by Meisner. This conclusion is supported by the fact that such a large discrepancy does not occur in the PMP#HT, PMP#T and PT estimates for kurtosis.

3. The scatter in the PMP#HT, PMP#T and PT estimates about the "true" indicate the difficulties which exist in choosing the end approximations. Their choice should be based on the behaviour of $s(y)$ in the neighbourhood of the origin. When such information is available it must be utilized. Unfortunately, in most applications, *a priori* information about the behaviour of $s(y)$ is non-existent.

4. Unless sufficient data is available, end approximations must be applied. As the difference between PMP#HT and PMP#T (without the use of an end approximation) indicates, some of the values of the functionals in Table 2 could be in error by more than 10%. Any decision to drop the application of end approximations will depend heavily on the not-

TABLE 1

SOME FUNCTIONALS OF IMPORTANCE IN APPLICATIONS

Functional	Notational Definition	Corresponding Data Functional
mean (first moment)	$m = M_1$	$\bar{m} = \pi/2D_{-1}$
variance (second moment about mean)	$\sigma^2 = M_2^2 - m^2$	$\bar{\sigma}^2 = 4\bar{m}D_1/\pi - \bar{m}^2$
standard deviation	$s = \sqrt{\sigma^2}$	$\bar{s} = \sqrt{\bar{\sigma}^2}$
mean (spherical) surface area (proportional to second moment)	$4\pi M_2$	$16\bar{m}D_1$
skewness	$\bar{M}_3/(\sigma^2)^{3/2}$	$\{3\bar{m}D_2/2 - 3\bar{m}\bar{\sigma}^2 - \bar{m}^3\} / (\bar{\sigma}^2)^{3/2}$
	$(\bar{M}_3 = M_3 - 3m\sigma^2 - m^3)$	
mean (spherical) volume (proportional to third moment)	$\frac{4\pi}{3}M_3$	$2\pi\bar{m}D_2$
kurtosis	$\bar{M}_4 / (\sigma^2)^2$	$\{16\bar{m}D_3/3\pi + 6\bar{m}^2\bar{\sigma}^2 + 3\bar{m}^4 - 6\bar{m}^2D_2\} / (\bar{\sigma}^2)^2$
	$(\bar{M}_4 = M_4 + 6m^2\sigma^2 + 3m^4 - 4mM_3)$	

TABLE 2

COMPARISON OF METHODS FOR MEISNER#1 DATA

	True	M	PMP#HT	PMP#T	PT	MDF
mean ($\times 10^{-3}$)	7.500	7.500	7.533	7.448	7.477	7.501
variance ($\times 10^{-6}$)	3.083		2.980	3.456	3.237	
standard deviation ($\times 10^{-3}$)	1.756	1.732	1.726	1.859	1.799	1.752
mean spherical surface area ($\times 10^{-4}$)	7.456		7.505	7.406	7.431	
skewness	2.909	3.031	3.347	2.280	2.561	3.079
mean spherical volume ($\times 10^{-7}$)	21.106		21.445	21.157	21.172	
kurtosis	11.740	9.233	11.353	10.371	10.332	7.944

TABLE 3

COMPARISON OF METHODS FOR MEISNER#2 DATA

	True	M	PMP#HT	PMP#T	PT	MDF
mean ($\times 10^{-3}$)	2.500	2.466	2.573	2.506	2.529	2.511
variance ($\times 10^{-6}$)	4.573		4.668	4.629	4.741	
standard deviation ($\times 10^{-3}$)	2.138	2.140	2.161	2.151	2.177	2.148
mean spherical surface area ($\times 10^{-4}$)	1.360		1.419	1.371	1.400	
skewness	3.001	2.964	3.035	3.036	3.049	3.036
mean spherical volume ($\times 10^{-7}$)	3.320		3.506	3.383	3.503	
kurtosis	15.364	12.197	15.668	15.775	15.951	12.626

TABLE 4

COMPARISON OF METHODS FOR MEISNER#3 DATA

	PMP#HT	PMP#T	PT
mean ($\times 10^{-3}$)	2.837	2.837	2.831
variance ($\times 10^{-6}$)	3.445	3.445	3.431
standard deviation ($\times 10^{-3}$)	1.856	1.856	1.852
mean spherical surface area ($\times 10^{-4}$)	1.444	1.444	1.438
skewness	2.570	2.570	2.574
mean spherical volume ($\times 10^{-7}$)	2.873	2.873	2.856
kurtosis	7.109	7.109	7.097

unreasonable assumptions that (a) the underlying distribution $s(y)$ is not pathological in the neighbourhood of zero or infinity, and (b) the resolution of the measurement process used is such that, if the amount of data were increased, the change in the values of the functionals being estimated is insignificant.

5. As Table 3 indicates, when PMP#HT, PMP#T and PT are applied to MEISNER#2, they consistently overestimate all the functionals of Table 1. For this data, this implies that the form of $s(y)$ in the neighbourhood of the origin defined by the lowest sampled data point is such that the HT and T end approximations are too crude. This is confirmed in part by Table 1 which shows for these data that the end approximations are relatively small compared with those for the MEISNER#2 data. In addition, the PMP#HT, PMP#T and PT estimates derived for the MEISNER#1 data do not consistently overestimate. In fact, except for variance and standard deviation, the PMP#HT overestimate while the PMP#T and PT underestimate, indicating that something between an HT and a T end approximation is required.

6. The conclusions of both 4 and 5 are consistent with the fact that a higher sampling rate has been used for MEISNER#2 (24 data points) than MEISNER#1 (16 data points).

Together, these conclusions represent strong evidence for *the use, where possible, if functionals evaluated on the data* $\{d_i\}$ *rather than the corresponding functionals evaluated using the direct strategy.*

In order to compare the effect of grouping sample cumulative data, the algorithms PMP#HT, PMP#T and PT were applied to MEISNER#3 and MEISNER#4 to compute the functionals given in Table 1. The results are listed in Tables 4 and 5 (respectively). (Because the cumulative data of MEISNER#3 is listed as a density (cf. Appendix B), PMP#HT, PMP#T and PT were applied directly to them.)

TABLE 5

COMPARISON OF METHODS FOR MEISNER#4 DATA

	M	PMP#HT	PMP#T	PT	MDF
mean ($\times 10^{-3}$)	2.033	2.848	2.848	2.875	2.825
variance ($\times 10^{-6}$)		3.501	3.501	3.501	
standard deviation ($\times 10^{-3}$)	1.826	1.871	1.871	1.871	1.878
mean spherical surface area ($\times 10^{-4}$)		1.459	1.459	1.478	
skewness	2.430	2.577	2.577	2.563	2.505
mean spherical volume ($\times 10^{-7}$)		2.927	2.927	2.963	
kurtosis	5.533	7.208	7.208	2.019	3.995

The estimates obtained from MEISNER#3 are consistently better than those obtained from MEISNER#4 in the following ways :

a. For each of the functionals listed, the scatter in the values of the different estimates derived from MEISNER#3 is very small, and the corresponding end approximations (as estimated as the difference between PMP#HT and PMP#T) are nelgigible.

b. The larger scatter is in the values derived from MEISNER#4. In particular, the lack of agreement of PT compared with PMP#HT and PMP#T.

c. The very poor agreement for MEISNER#4 between M and MDT. One would expect a better result for M if MEISNER#3 had been used instead of MEISNER#4 for its application.

d. Except for the standard deviation s, the poor agreement for MEISNER#4 of MDF compare with PMP#HT, PMP#T and PT.

In fact, these conclusions represent strong evidence for *the use of a strategy for the evaluation of linear functionals which is based on*

(i) *the use of a high rather than a low sampling rate,*

and

(ii) *the use of cumulative data.*

Additional advantages in favour of the use of the data functionals strategy with sample cumulative data are :

A. If, in addition to certain specified functionals $m_{\theta_j}(u(y))$, $j = 1,2,\ldots,J$, an accurate solution of (2.1) is required with respect to given data $\{d_i\}$, then a comparison of the specified functionals $m_{\theta_j}(u(y))$ with the corresponding data functional $m_{\phi_j}(s(y))$ can be used as an independent measure to compare the quality of possible numerical solutions.

B. Via the use of the more sophisticated methods of the type discussed in Anderssen [1], [2] and Anderssen and Jakeman [3], numerical approximations to u(y) can be generated which, when used to evaluate $m_\theta(u(y))$, yield values which compare favourably with those generated when the corresponding $m_\phi(s(y))$ is evaluated using the data $\{d_i\}$. It will then be necessary to base the choice between the two strategies on computational complexit considerations. Where such considerations are relevant, they will tend to favour the us of the data functionals strategy.

C. The amount of data may be insufficient to allow a suitably accurate numerical soluti of (2.1) to be determined and hence allow the direct method to be implemented, although i may be sufficient to allow an accurate estimate of some of the simpler functionals to be computed using the data functional strategy. The use of sample cumulative (non-grouped) data is essential from this point of view.

§4. Extension of the Data Functional Strategy to General Integral Equations

For general linear operator equations

$$\underset{\sim}{A}u = s \ , \quad A : \underline{\underline{D}}(\underset{\sim}{A}) \to \underline{\underline{R}}(\underset{\sim}{A}) \ , \tag{4.1}$$

where $\underline{\underline{D}}(\underset{\sim}{A})$ and $\underline{\underline{R}}(\underset{\sim}{A})$ denote respectively the domain and range of the operator $\underset{\sim}{A}$, the general strategy for the evaluation of linear functionals $m_\theta(u)$ defined on the solution u as a linear functional $m_\phi(s)$ defined on the data s is easily stated and proved; but, because of the underlying regularity conditions, can be difficult to implement.

THEOREM 4.1 *Let* $\underset{\sim}{A}^*$ *denote the adjoint of* $\underset{\sim}{A}$ *with respect to the* L_2*-inner product*

$$(u,v) = \int_0^a uv dx \ . \tag{4.2}$$

If the known θ *which defines* $m_\theta(u)$ *is contained in* $\underline{\underline{R}}(A^*)$, *then the required* ϕ *which defines* $m_\phi(s)$ *is given by*

$$\underset{\sim}{A}^*\phi = \theta \ . \tag{4.3}$$

Proof

$$m_\theta(u) = (\theta,u) = (\underset{\sim}{A}^*\phi,u) = (\phi,\underset{\sim}{A}u) = (\phi,s) = m_\phi(s) \ . \quad \#$$

Golberg [5] has used this result to explain the use of the data functional strategy in a number of specific applications areas (viz., Abel-type integral equations, and reverse flow theorems), but has not explored its more general applicability. This is limited by the requirement that θ be contained in $R(A^*)$.

For general first kind Volterra equations of the form

$$\underset{\sim}{K}u = \int_0^y K(y,x)u(x)dx = s(y) \ , \quad 0 \le y \le a \ , \tag{4.4}$$

conditions which guarantee that $\theta \in \underline{\underline{R}}(K^*)$ are given by

THEOREM 4.2 *For the first kind Volterra equation (4.4), assume that the following conditions are satisfied :*

(a) $K(y,y) \neq 0 \ , \quad 0 \le y \le a$,

(b) $\partial K(x,y) / \partial y$ *is continuous for* $0 \le x,y \le a$.

Then, $\theta \in \underline{\underline{R}}(K^*)$ *if* $d\theta(x)/dx$ *is continuous for* $0 \le x \le a$. *In addition, the solution of* $\underset{\sim}{K}^*\phi = \theta$ *is defined by*

$$- K(y,y)\phi(y) + \int_y^a \frac{\partial K(x,y)}{\partial y}\phi(x)dx = \theta'(y) , \quad 0 \le y \le a . \tag{4.5}$$

Proof From (4.4), it follows that $\underset{\sim}{K}^*\phi = \theta$ takes the form

$$\underset{\sim}{K}^*u = \int_y^a K(x,y)u(x)dx = \theta(y) , \quad 0 \le y \le a .$$

Differentiation with respect to y yields (4.5). If θ is such that $d\theta(x)/dx$ is continuous for $0 \le x \le a$, then conditions (a) and (b) guarantee that (4.5) uniquely determines ϕ (cf., Schmeidler (1950), Theorem 52, page 246). #

COROLLARY 4.1 *Assume that conditions* (a) *and* (b) *of Theorem* 4.2 *hold. When* $\theta(x) = x^p$, *the corresponding data functionals* $m_\phi(s(x))$ *exist for* $p \ge 1$.

Proof An immediate consequence of Theorem 4.2, since $d\theta(x)/dx$ is continuous for $0 \le x \le a$ if $p \ge 1$. When $p = 0$, $d\theta(x)/dx \equiv 0$, which yields $\phi(y) \equiv 0$. #

Specifying general conditions under which $\theta \in \underline{\underline{R}}(A^*)$ for first kind Fredholm equations is more difficult. In fact, the requirement that $\theta \in \underline{\underline{R}}(\underset{\sim}{A}^*)$ imposes quite strong constraints on the extension and applicability of the data functional strategy to such integral equations as the following theorem indicates.

THEOREM 4.3 *For the first kind Fredholm integral equation*

$$\int_0^a k(x,y)u(y)dy = s(x) , \quad 0 \le x \le a , \tag{4.6}$$

where $k(x,y)$ *satisfies the following conditions*

(a) *for fixed* y, $0 \le y \le a$, $k(x,y) \in C^{j+1}[0,a]$, $j \ge 0$,

and

(b) $k^{(j)}(x,y) = \partial^j k(x,y)/\partial x^j$ *is a non-degenerate Hilbert Schmidt kernel,*

the data functional $m_\phi(s(x))$ *corresponding to* $m_\theta(u(y))$, $\theta = y^j$, *does not exist.*

Proof Assume the contrary; namely, that there exists a non-zero $\phi(x)$ for which the required $m_\phi(s(x))$ is defined. Then the defining relationship for $\phi(x)$ is

$$\int_0^a k(y,x)\phi(x)dx = y^j , \quad j \ge 0 , \quad 0 \le y \le a . \tag{4.7}$$

Differentiating (4.7) j+1-times yields

$$\underset{\sim}{K}\phi = \int_0^a k^{(j+1)}(y,x)\phi(x)dx \equiv 0 , \quad 0 \le y \le a . \tag{4.8}$$

Condition (b) implies that (Mercer's Theorem - cf. Zabreyko et al (1975), Chapter III, §5)

$$k^{(j+1)}(y,x) = \sum_{i=0}^{\infty} \sigma_i \phi_i(y)\psi_i(x) ,$$

where $\{\phi_i(y)\}$ and $\{\psi_i(x)\}$ denote complete orthonormal eigenfunctions of $\underset{\sim}{K}^*\underset{\sim}{K}$ and $\underset{\sim}{K}\underset{\sim}{K}^*$, respectively, and $\sigma_i = +(\lambda_i^2)$, where the λ_i^2 denote the eigenvalues of $\underset{\sim}{K}^*\underset{\sim}{K}$ (and, of course, $\underset{\sim}{K}\underset{\sim}{K}^*$). Substitution of this representation for $k^{(j+1)}(y,x)$ in (4.8) yields

$$\sum_{i=0}^{\infty} \lambda_i \phi_i(y) \int_0^a \psi_i(x)\phi(x)dx = 0$$

which implies that $(\psi_i(x),\phi(x)) = 0$, $i = 1,2,\ldots,$ and hence that $\phi \equiv 0$, $0 \le x \le Y$, which yields the desired contradiction. #

As an immediate consequence, we obtain

COROLLARY 4.2 *If conditions* (a) *and* (b) *of Theorem 4.3 hold for* $j = 0,1,2,\ldots,n$, *then the data functional* $m_\phi(s(x))$ *corresponding to* $m_\theta(u(y))$, $\theta(y) = p_n(y)$, *where* $p_n(y)$ *denotes a polynomial of degree* $\le n$, *does not exist.*

Example 4.1 An example of a first kind Fredholm equation for which conditions (a) and (b) of Theorem 4.3 are satisfied is given by the limb-darkening equation examined by Holt and Jupp [7] and describing the variation in emergent intensity across the solar disc from the centre of the limb; viz.,

$$\mu I(\mu) = \int_0^\infty \exp(-t/\mu)S(t)dt , \qquad 0 \le \mu < \infty , \tag{4.9}$$

where $I(\mu)$ denotes the emergent intensity at an angle $\theta = \cos^{-1}\mu$ to the normal from a plane parallel, semi-infinite atmosphere, $S(t)$ the source function and t the monochromatic optical depth. #

Note 4.1 Theorems 4.2 and 4.3 do not imply that functionals $m_\theta(u(y))$, $\theta(y) = y^p$, for which corresponding data functionals $m_\phi(s(y))$ do not exist, cannot be estimated in some direc manner from the data $s(y)$. We give two examples. For the limb-darkening equation (4.9), the use of the standard moment generating function argument yields

$$\sum_{i=0}^{\infty} y^i(0) \frac{\mu^i}{i!} = \sum_{i=0}^{\infty} \frac{(-1)^i}{i!} \frac{1}{\mu^i} \int_0^\infty t^i S(t)dt , \qquad y(\mu) = \mu I(\mu) ,$$

for which the only solution is

$$\int_0^\infty S(t)dt = \lim_{\mu\to 0} \mu I(\mu) .$$

For the Abel-type integral equation (2.1), we have already seen that

$$m = \int_0^a xu(x)dx = \frac{\pi}{2}\left(\int_0^a \frac{s(x)}{x}dx\right)^{-1} ,$$

which is an immediate consequence of (2.1) if one first multiplies through by (m/y) and then integrates from 0 to a with respect to y . #

Appendix A

Product Integration Algorithms for Data Functionals

For the types of applications under consideration, the observed discrete data $\{d_i\}$ defined by (1.5) will correspond to either a sample density $d_n(x)$ or a sample cumulative $C_n(x)$ of the required size distribution $s(x)$ (cf. Figures 3 and 4). As has already been noted in §1, the approximation of $s(x)$ derived from $C_n(x)$ must be differentiated before it can be used to approximate $m_\phi(s)$, while the approximation derived from $d_n(x)$ can be used directly. Consequently, an examination of algorithms for linear functionals divides naturally into two parts; namely, algorithms for sample density data $d_n(x)$, and algorithms for sample cumulative data $C_n(x)$.

The steps in the sample cumulative are assumed to occur at the points x_i of (1.5); whereas the steps in the sample density occur at the boundaries of the size classes which are usually defined to have the points x_i as their mid-points (cf. Figures 3 and 4). Consequently, the size classes for the sample density are defined by the partition.

$$P_n = \{\bar{x}_i\} = \left\{\bar{x}_i ;\ 0 \le \bar{x}_0 < \bar{x}_1 < \ldots < \bar{x}_n ,\quad x_i = \frac{\bar{x}_{i-1}+\bar{x}_i}{2}\right\} . \tag{A.1}$$

§A.1 Algorithms for Linear Functionals Using Sample Density Data

When the given data $\{d_i\}$ corresponds to a sample density, the following strategy is used for the construction of algorithms for the data functionals $m_\phi(s)$:

Approximate $d_n(x)$ *by* $\tilde{d}_n(x)$ *under the proviso that, with respect to a given* $\phi(x)$ *, the functional* $m_\phi(\tilde{d}_n(x))$ *can be evaluated analytically.*

Clearly, for arbitrary $\phi(x)$, there exists no unique $\tilde{d}_n(x)$ such that it will always be possible to evaluate $m_\phi(\tilde{d}_n(x))$ analytically. Therefore, since the aim is to make a comparison with the functionals computed in Meisner [13], we restrict attention to the moment data functionals

$$D_p(s) = \int_0^a x^p s(x)dx , \quad a < \infty , \quad p \ge -1 . \tag{A.2}$$

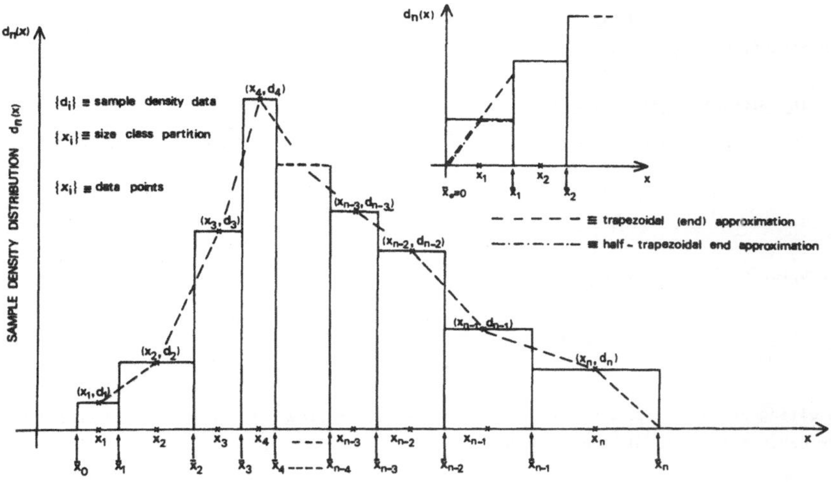

FIGURE 3

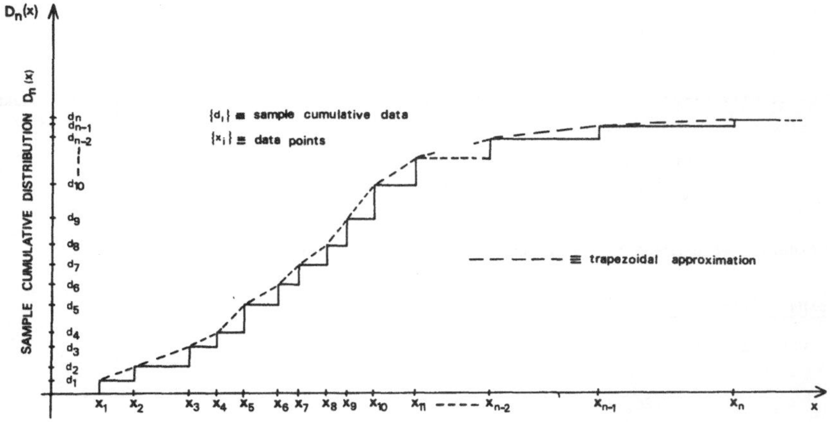

FIGURE 4

With respect to the given partition P_n , the form of the discretization applied to $D_p(s)$ will depend on the form of $\tilde{d}_n(x)$. We examine two possibilities : the mid-point and the trapezoidal approximations.

A.1.1 The Mid-Point Approximation

When the sample density $d_n(x)$ is itself taken to define the approximation $\tilde{d}_n(x)$, namely

$$\tilde{d}_n(x) = d_i \ , \quad \bar{x}_{i-1} \le x < \bar{x}_i \ , \quad i = 1,2,\ldots,n \ , \tag{A.3}$$

the appropriate discretization is given by

$$D_p(d_n(x)) = \sum_{i=0}^{n-1} \int_{\bar{x}_i}^{\bar{x}_{i+1}} x^p d_n(x)dx \ . \tag{A.4}$$

The resulting procedures derived for evaluating $D_p(s)$ on the data $\{d_i\}$ are often referred to as product mid-point integration formulas.

In fact, (A.4) yields for $p > -1$

$$D_p(d_n(x)) = \left\{\sum_{i=0}^{n-1} d_{i+1}\left(\bar{x}_{i+1}^{p+1} - \bar{x}_i^{p+1}\right)\right\} / (p+1) \ , \quad p > -1 \ ; \tag{A.5}$$

while for $p = -1$ with $\bar{x}_0 > 0$, it yields

$$D_{-1}(d_n(x)) = \sum_{i=0}^{n-1} d_{i+1}\ln(\bar{x}_{i+1}/\bar{x}_i) \ , \quad \bar{x}_0 > 0 \ . \tag{A.6}$$

The only difficulty associated with evaluating (A.4) analytically arises when $\phi(x) = x^{-1}$ and $\bar{x}_0 = 0$, since

$$\int_0^{\bar{x}_1} \{d_n(x)/x\}dx \ , \quad d_n(x) = d_1 \ , \quad 0 \le x \le \bar{x}_1 \ ,$$

is not defined. We examine two modifications which circumvent this difficulty :

(i) The Half-Trapezoidal End Approximation (cf. Figure 3)

If, when $\bar{x}_0 = 0$, the half trapezoidal end approximation

$$\tilde{d}_n(x) = \begin{cases} 2d_1 x/\bar{x}_1 \ , & 0 \le x \le \bar{x}_1/2 \ , \\ d_1 \ , & \bar{x}_1/2 \le x < \bar{x}_1 \ , \end{cases} \qquad \tilde{d}_n(x) = d_n(x) \ , \ \bar{x}_1 \le x \le \bar{x}_n \ ,$$

is used in place of $\tilde{d}_n(x) = d_n(x)$, then the estimates (A.5) and (A.6) are replaced by

$$D_p(\tilde{d}_n(x)) = d_1(\bar{x}_1/2)^{p+1} / (p+2) + d_1\left(\bar{x}_1^{p+1} - (\bar{x}_1/2)^{p+1}\right) / (p+1)$$

$$+ \left\{\sum_{i=1}^{n-1} d_{i+1}\left(\bar{x}_{i+1}^{p+1} - \bar{x}_i^{p+1}\right)\right\} / (p+1) \ , \quad p > -1 \ , \qquad (A.7)$$

$$D_{-1}(\tilde{d}_n(x)) = (1+\ell n\ 2)d_1 + \sum_{i=1}^{n-1} d_{i+1}\ell n(\bar{x}_{i+1}/\bar{x}_i) \ . \qquad (A.8)$$

(ii) <u>The Trapezoidal End Approximation</u> (cf. Figure 3)

If, when $\bar{x}_0 = 0$, the trapezoidal end approximation

$$\tilde{d}_n(x) = 2d_1 x/\bar{x}_1 \ , \ 0 \le x \le \bar{x}_1 \ , \ \tilde{d}_n(x) = d_n(x), \ \bar{x}_1 \le x \le \bar{x}_n \ ,$$

is used in place of $\tilde{d}_n(x) = d_n(x)$, then the estimates (A.5) and (A.6) are replaced by

$$D_p(\tilde{d}_n(x)) = 2d\bar{x}_1^{p+1} / (p+2) + \left\{\sum_{i=1}^{n-1} d_{i+1}\left(\bar{x}_{i+1}^{p+1} - \bar{x}_i^{p+1}\right)\right\} / (p+1) \ , \ p > -1 \ , \qquad (A.9)$$

$$D_{-1}(\tilde{d}_n(x)) = 2d_1 + \sum_{i=1}^{n-1} d_{i+1}\ell n(\bar{x}_{i+1}/\bar{x}_i) \ . \qquad (A.10)$$

<u>Notation</u>. For notational convenience, we shall denote the use of (A.7) and (A.8), based on the half-trapezoidal end approximation, by PMP#HT; and the use of (A.9) and (A.10), based on the trapezoidal end approximation, by PMP#T.

§A.1.2. <u>The Trapezoidal Approximation</u>

When, on the interval $x_1 \le x \le x_n$, $\tilde{d}_n(x)$ is defined to be the trapezoidal approximation

$$\tilde{d}_n(x) = \frac{x_{i+1}d_i - x_i d_{i+1}}{x_{i+1} - x_i} + \frac{d_{i+1} - d_i}{x_{i+1} - x_i}x, \ x_i \le x \le x_{i+1} \ , \ i = 1,2,\dots,n-1 \ , \qquad (A.11)$$

the appropriate discretization to apply is

$$D_p(\tilde{d}_n(x)) = \int_{\bar{x}_0}^{x_1} x^p\tilde{d}_n(x)dx + \sum_{i=1}^{n-1}\int_{x_i}^{x_{i+1}} x^p\tilde{d}_n(y)dy + \int_{x_n}^{\bar{x}_n} x^p\tilde{d}_n(x)dx \ . \qquad (A.12)$$

The resulting procedure derived for evaluating $D_p(s)$ on the data $\{d_i\}$ are often referred to as *product trapezoidal formulas* and take the form

$$D_p(\tilde{d}_n(x)) = K_0^{(p)} + K_n^{(p)} + \left\{\sum_{i=1}^{n-1} \frac{x_{i+1}d_i - x_i d_{i+1}}{x_{i+1} - x_i}\left(x_{i+1}^{p+1} - x_i^{p+1}\right)\right\} / (p+1)$$

$$+ \left\{\sum_{i=1}^{n-1} \frac{d_{i+1} - d_i}{x_{i+1} - x_i}\left(x_{i+1}^{p+2} - x_i^{p+2}\right)\right\} / (p+2) \ , \quad p > -1 \ , \qquad (A.13)$$

$$D_{-1}(\tilde{d}_n(x)) = K_0^{(-1)} + K_n^{(-1)} + \sum_{i=1}^{n-1} \frac{x_{i+1}d_i - x_i d_{i+1}}{x_{i+1}-x_i} \ln(x_{i+1}/x_i) + d_n - d_1 , \qquad (A.14)$$

where

$$K_0^{(p)} = \int_{\bar{x}_0}^{x_1} x^p \tilde{d}_n(x)dx , \quad K_n = \int_{x_n}^{\bar{x}_n} x^p \tilde{d}_n(x)dx . \qquad (A.15)$$

Thus, the complete specification of these estimates depends on how $\tilde{d}_n(x)$ is defined on the intervals $\bar{x}_0 \le x < x_1$ and $x_n \le x \le \bar{x}_n$. If the trapezoidal end approximations are used, namely,

$$\tilde{d}_n(x) = d_1(x-\bar{x}_0) / (x-\bar{x}_0) , \quad \bar{x}_0 \le x < x_1 ,$$

$$\tilde{d}_n(x) = d_n(x-\bar{x}_n) / (x_n-\bar{x}_n) , \quad x_n \le x \le \bar{x}_n ,$$

then

$$K_0^{(p)} = d_1 \begin{cases} \left[(x_1^{p+2} - \bar{x}_0^{p+2}) / (p+2) - \bar{x}_0(x_1^{p+1} - \bar{x}_0^{p+1}) / (p+1)\right] / (x_1-\bar{x}_0) , & p > -1 , \ \bar{x}_0 \ge 0 , \\ [1 - \bar{x}_0 \ln(x_1/\bar{x}_0) / (x_1-\bar{x}_0)] , & p = -1 , \ \bar{x}_0 > 0 , \\ 1 , & p = -1 , \ \bar{x}_0 = 0 \end{cases} \qquad (A.16)$$

and

$$K_0^{(p)} = d_n \begin{cases} \left[(\bar{x}_n^{p+2} - x_n^{p+2}) / (p+2) - \bar{x}_n(\bar{x}_n^{p+1} - x_n^{p+1}) / (p+1)\right] / (x_n-\bar{x}_n) , & p > -1 , \\ [-1 - \bar{x}_n \ln(\bar{x}_n/x_n) / (x_n-\bar{x}_n)] , & p = -1 . \end{cases} \qquad (A.17)$$

Notation. For notational convenience, we shall denote the use of (A.13) - (A.17) by PT.

Caveat. Depending on the form chosen for the end approximations, the area weighting of the frequency counts on size classes must be adjusted accordingly.

§A.2 Algorithms for Linear Functionals Using Sample Cumulative Data

When the given data corresponds to a sample cumulative, the following strategy is used for the construction of algorithms for the data functionals $m_\phi(s)$:

Approximate $C_n(x)$ *by* $\tilde{C}_n(x)$ *under the proviso that, with respect to a given* $\phi(x)$ *, the functional*

$$m_\phi(\tilde{C}_n'(x)) = \int_{\bar{x}_0}^{\bar{x}_n} \phi(x)\tilde{C}_n'(x)dx , \quad \tilde{C}_n'(x) = d\tilde{C}_n(x)/dx , \qquad (A.18)$$

can be evaluated analytically.

For reasons equivalent to those given in §A.1, we restrict attention to the moment functionals (A.2). We examine the use of the trapezoidal approximation.

§A.2.1. The Trapezoidal Approximation

When, on the interval $x_1 \le x \le x_n$, $\tilde{C}_n(x)$ is defined to be the trapezoidal approximation

$$\tilde{C}_n(x) = \frac{x_{i+1}d_i - x_i d_{i+1}}{x_{i+1}-x_i} + \frac{d_{i+1}-d_i}{x_{i+1}-x_i}\, x \ , \ x_i \le x < x_{i+1} \ , \ i = 1,2,\dots,n-1 \ , \qquad (A.18)$$

the appropriate discretization to apply is

$$D_p(\tilde{C}_n'(x)) = \sum_{i=1}^{n-1} \int_{x_i}^{x_{i+1}} x^p \tilde{C}_n'(x)\,dx \ . \qquad (A.19)$$

Because the derivative of $\tilde{C}_n(x)$ is used in (A.19), the approximation generated by $D_p(\tilde{C}_n'(x))$ will be similar to the product mid-point approximation generated by $D_p(d_n(x))$, and will take the form (cf. (A.5) and (A.6))

$$D_p(\tilde{C}_n'(x)) = \left\{\sum_{i=1}^{n-1} \frac{d_{i+1}-d_i}{x_{i+1}-x_i} \left(x_{i+1}^{p+1} - x_i^{p+1}\right)\right\} / \ (p+1) \ , \ p > -1 \ , \qquad (A.20)$$

$$D_{-1}(\tilde{C}_n'(x)) = \sum_{i=1}^{n-1} \frac{d_{i+1}-d_i}{x_{i+1}-x_i}\, \ell n(x_{i+1}/x_i) \ . \qquad (A.21)$$

Depending on circumstances, end corrections could be applied. Since they are not the major concern of this paper, only the estimates (A.20) and (A.21) will be examined here. For details about methods which allow for the natural truncation inherent in discrete data, the interested reader is referred to Nicholson [14] and Jakeman and Anderssen [11].

In addition, the sample density generated by $\tilde{C}_n'(x)$ may not satisfy

$$\int_{x_1}^{x_n} \tilde{C}_n'(x)\,dx = 1 \ .$$

Though various strategies can be applied to correct for this discrepancy, only the weighting of the estimates by

$$1 \ / \left\{\int_{x_1}^{x_n} \tilde{C}_n'(x)\,dx\right\} \qquad (A.22)$$

will be considered here.

Notation. For notational convenience, we shall denote the use of (A.20) and (A.21) along with the weighting (A.22) by PTC.

APPENDIX B

THE DATA ANALYSED BY MEISNER

Since a major aim of the paper is a comparison with the functionals given in Meisner [13] we give full details about the data from which they were derived. Of the four size distributions he uses, the first two (namely, MEISNER#1 and MEISNER#2) are synthetic sample density distributions, the third (namely, MEISNER#3) is an observed sample cumulative and the fourth (namely, MEISNER#4) is a grouped density derived from the observed sample cumulative (Figure 2). The actual values along with the corresponding size classes are listed in the Table.

The details are :

<u>MEISNER#1</u>. For this data, the true distribution for the sphere diameters was taken to be a probability density function compounded from two uniform densities :

$$u(x) = \begin{cases} \lambda(\beta-\alpha)^{-1}, & \alpha \le x < \beta, \\ (1-\lambda)(\gamma-\beta)^{-1}, & \beta \le x \le \gamma, \\ 0, & x < \alpha, \quad x > \gamma. \end{cases}$$

A simple integration of (1.4) with $\mu = \frac{1}{2}$ then yields the true distribution for the circle diameters $s(x)$ (see Example (1) of Meisner [13] for the lengthy formulas). The actual data, derived from the latter, is a grouped distribution of observed diameters with 16 equal size classes as shown in the Table.

<u>MEISNER#2</u>. For this data, the true distribution for the sphere diameters was taken to be

$$u(x) = 0.9\,\lambda_1^{-2}\, x \exp(-0.5\lambda_1^{-2}x^2) + 0.1\,\lambda_2^{-2}\, x \exp(-0.5\lambda_2^{-2}x^2),$$

from which it follows that

$$s(x) = \{0.9\lambda_1^{-1}\, x \exp(-0.5\lambda_1^{-2}x^2) + 0.1\,\lambda_2^{-1}\, x \exp(-0.5\lambda_2^{-2}x^2)\} / m$$

where, as in (2.2), m denotes the mean sphere diameter. The actual data, derived from the latter, is a grouped distribution of observed diameter with 24 size classes as shown in the Table.

<u>MEISNER#3</u>. This data consists of a sample cumulative of observed circle diameters determined from a photomicrographic analysis of a plane section of high-impact polystyrone (for further details, see Example (3) in Meisner [13]).

<u>MEISNER#4</u>. This data was derived from MEISNER#3 by grouping into 16 size classes. The details are given in the Table.

TABLE

OBSERVED CIRCLE DIAMETERS OF MEISNER [13]

Class Number i	MEISNER#1		MEISNER#2		MEISNER#3		MEISNER#3 (continued)		MEISNER#4	
	Class Limits	Frequency Counts	Class Limits	Frequency Counts	Class Limits	Frequency Counts	Class Limits	Frequency Counts	Class Limits	Frequency Counts
1	0- 1	9.25541	0 - 0.5	35.60922	0.59-0.63	1	2.82- 3.01	37	0. - 0.67	6
2	1- 2	28.35956	0.5 - 0.6	15.24979	0.63-0.67	5	3.01- 3.21	19	0.67- 0.82	42
3	2- 3	49.46609	0.6 - 0.72	21.24979	0.67-0.72	14	3.21- 3.43	27	0.82- 0.99	148
4	3- 4	74.89306	0.72 - 0.864	29.49632	0.72-0.76	13	3.43- 3.66	24	0.99- 1.21	202
5	4- 5	109.98979	0.864 - 1.0368	40.29075	0.76-0.82	15	3.66- 3.91	40	1.21- 1.47	149
6	5- 6	179.82480	1.0368- 1.2442	53.78319	0.82-0.87	49	3.91- 4.17	42	1.47- 1.79	82
7	6- 7	267.73803	1.2442- 1.4930	69.47048	0.87-0.93	44	4.17- 4.45	30	1.79- 2.17	64
8	7- 8	165.95892	1.4930- 1.7916	85.62947	0.93-0.99	55	4.45- 4.75	41	2.17- 2.64	67
9	8- 9	17.64108	1.7916- 2.1499	98.79743	0.99-1.06	61	4.75- 5.07	32	2.64- 3.21	83
10	9-10	17.58730	2.1499- 2.5799	103.98828	1.06-1.13	78	5.07- 5.41	31	3.21- 3.91	91
11	10-11	17.19448	2.5799- 3.0959	96.76285	1.13-1.21	63	5.41- 5.78	31	3.91- 4.75	113
12	11-12	16.42860	3.0959- 3.7150	77.44198	1.21-1.29	53	5.78- 6.17	44	4.75- 5.78	94
13	12-13	15.22473	3.7150- 4.4580	54.18517	1.29-1.38	51	6.17- 6.58	49	5.78- 7.03	124
14	13-14	13.45663	4.4580- 5.3497	38.55905	1.38-1.47	45	6.58- 7.03	31	7.03- 8.55	83
15	14-15	10.83521	5.3497- 6.4196	34.73917	1.47-1.57	31	7.03- 7.50	29	8.55-10.39	41
16	15-16	6.14631	6.4196- 7.7035	36.69444	1.57-1.67	27	7.50- 8.00	30	10.39-11.84	7
17			7.7035- 9.2442	36.99452	1.67-1.79	24	8.00- 8.55	24		
18			9.2442-11.093	32.21534	1.79-1.91	21	8.55- 9.12	15		
19			11.093 -13.312	22.59922	1.91-2.03	26	9.12- 9.73	17		
20			13.312 -15.974	11.66778	2.03-2.17	17	9.73-10.39	9		
21			15.974 -19.169	3.91199	2.17-2.32	31	10.39-11.09	4		
22			19.169 -23.003	0.71702	2.32-2.48	20	11.09-11.84	3		
23			23.003 -27.603	0.05668	2.48-2.64	16				
24			27.603 -33.124	0.00139	2.64-2.82	27				
TOTAL	10^{-3}mm	1000	10^{-3}mm	1000	10^{-3}mm		10^{-3}mm	1396	10^{-3}mm	1396

REFERENCES

[1] R.S. Anderssen (1976), Stable procedures for the inversion of Abel's equation, *JIMA* 17 (1967), 329-342.

[2] R.S. Anderssen (1977), Application and Numerical Solution of Abel-type Integral Equations, *MRC Technical Summary Report* #1787, University of Wisconsin-Madison, September, 1977, pp.45.

[3] R.S. Anderssen and A.J. Jakeman, (1975a), Product integration for functionals of particle size distributions, *Utilitas Mathematica* 8 (1975), 111-126.

[4] R.S. Anderssen and A.J. Jakeman, (1975b), Abel type integral equations in stereology. II. Computational methods of solution and the random spheres approximation, *J. Micros.* 105 (1975), 135-153.

[5] M.A. Golberg, A method of adjoints for solving some ill posed equations of the first kind, preprint.

[6] J.E. Hilliard (1968) Direct determination of the moments of the size distribution of particles in an opaque field, *Trans. Metal. Soc. A.IMF* 242 (1968), 1373-1380.

[7] J.N. Holt and D.L.B. Jupp, (1978) Free-knot spline inversion of a Fredholm integral equation in astrophysics, *JIMA* 21 (1978), 429-443.

[8] E.D. Hyam and J. Nutting (1956), The tempering of plain carbon steels, *J. Iron and Steel Inst.* 148 (1956), 148-165.

[9] A.J. Jakeman (1975), *Numerical Inversion of Abel Type Integral Equations in Stereology*, Ph.D. Thesis, Australian National University, August, 1975.

[10] A.J. Jakeman (1976), *Numerical Inversion of a Second kind Singular Volterra Equation - The Thin Section Equation of Stereology*, Unpublished manuscript.

[11] A.J. Jakeman and R.S. Anderssen, (1975), Abel type integral equations in stereology. I. General discussion, *J. Micros.* 105 (1975), 121-133.

[12] A.J. Jakeman and R.L. Scheaffer (1977) On the properties of product integration estimators for linear functionals of particle size distributions,

[13] J. Meisner (1967) Estimation of the distribution of diameters of spherical particles from a given grouped distribution of diameters of observed circles formed by a plane section, *Statistica Neerlandica* 21 (1967), 11-30.

[14] W.L. Nicholson (1970) Estimation of linear properties of particle size distributions, *Biometrika* 57 (1970), 273-298.

[15] R.E. Miles and P. Davy (1977) On the choice of quadrats in stereology, *J. Micros.* (1977), 110, 27-44.

[16] R.E. Miles and P. Davy (1976) Precise and general conditions for the validity of a comprehensive set of stereological fundamental formulae, *J. Micros.*(1967) 107, 211-226.

[17] E.E. Underwood (1970) *Quantitative Stereology*, Addison-Wesley, Mass. 1970.

[18] G.S. Watson (1971) Estimating functionals of particle size distributions, *Biometrika* 58 (1971), 483-490.

[19] S.D. Wicksell (1925) The corpuscle problem, *Biometrika* 17 (1925), 84-99.

SYNTHESIS OF A SHAPED—BEAM REFLECTOR ANTENNA

Diet I. Ostry
Division of Radiophysics, CSIRO, Sydney, N.S.W.

Abstract

The role of integral equations in the analysis and design of microwave reflector antennas is briefly discussed. The synthesis problem for reflector antennas is formulated as a nonlinear Fredholm integral equation of the first kind and an example of its numerical solution using Newton's method is presented.

1. Introduction

Integral equations occur in both the analysis and synthesis of microwave reflector antennas. The determination of the electromagnetic field scattered by a reflector is a special case of the general scattering problem for the vector wave equation and leads to linear first and second kind Fredholm integral equations with Green's function kernels.

Reflector synthesis however is an application of the inverse scattering problem and, if electrically small reflectors are demanded, requires the solution of a nonlinear first kind Fredholm equation.

The principal aim of this paper is to present an example of the numerical solution of this nonlinear integral equation by a generalised Newton iteration. However, some background material on the scattering problem as applied to reflector antenna analysis is included in section 2 as an example of the integral equation formulation of exterior boundary value problems

2. THE SCATTERING PROBLEM

The scattering problem, that is the determination of the electromagnetic field in a medium containing a body subject to known excitation, is a fundamental problem in electromagnetism. When applied to reflector antennas, the scattering analysis is considerably simplified by assuming that the medium is loss-free, linear, homogeneous and isotropic and that the reflector is made of a perfectly conducting material. These assumptions are made here. Detailed discussions of the more general case may be found in [1], [2].

A typical configuration is shown in Figure 1. The electromagnetic field satisfies Maxwell's equations everywhere in the medium V. If harmonic time dependence $e^{i\omega t}$ is assumed but suppressed, these equations are

$$\nabla \times \underset{\sim}{E}(\underset{\sim}{r}) + i\omega\mu\underset{\sim}{H}(\underset{\sim}{r}) = 0 \tag{2.1}$$

$$\nabla \times \underset{\sim}{H}(\underset{\sim}{r}) - i\omega\varepsilon\underset{\sim}{E}(\underset{\sim}{r}) = \underset{\sim}{J}(\underset{\sim}{r}) . \tag{2.2}$$

$\underset{\sim}{E}$ and $\underset{\sim}{H}$ are the electric and magnetic fields, $\underset{\sim}{J}$ is the electric current density, and μ and ε are the permeability and permittivity of the medium. In addition, the conservation of charge gives

$$\nabla \cdot \underset{\sim}{J}(\underset{\sim}{r}) = -i\omega\rho(\underset{\sim}{r}) \tag{2.3}$$

where ρ is the charge density.

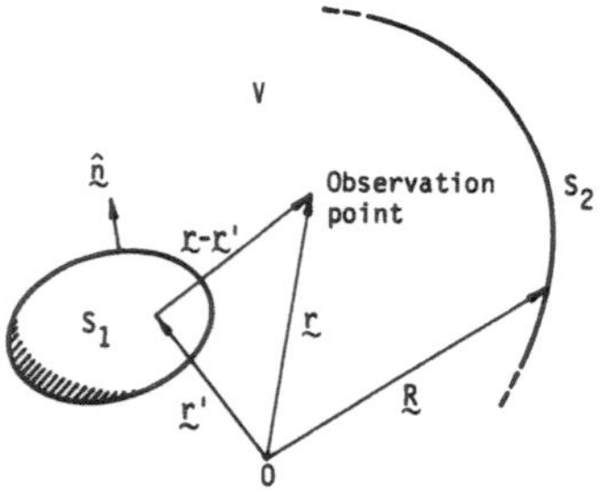

Fig. 1 - Geometry of the scattering problem.

If the medium has the properties mentioned above, μ and ε are real scalar constants then cross substitution of equations (2.1) and (2.2) gives the vector Helmholtz wave equations

$$\nabla \times \nabla \times \underset{\sim}{E}(\underset{\sim}{r}) - k^2\underset{\sim}{E}(\underset{\sim}{r}) = -i\omega\mu\underset{\sim}{J}(\underset{\sim}{r}) \tag{2.4}$$

$$\nabla \times \nabla \times \underset{\sim}{H}(\underset{\sim}{r}) - k^2\underset{\sim}{H}(\underset{\sim}{r}) = \nabla \times \underset{\sim}{J}(\underset{\sim}{r}) \tag{2.5}$$

where $k^2 = \omega^2\mu\epsilon$ is the square of the propagation constant. The tangential component of the electric field on the surface of a perfect conductor is zero so the boundary condition on the surface of the reflector (S_1) is

$$\hat{\underset{\sim}{n}} \times \underset{\sim}{E}(\underset{\sim}{r}') = 0 \tag{2.6}$$

where $\hat{\underset{\sim}{n}}$ is the outward normal to S_1 and a prime will be used to indicate a reference to points on S_1. If the fields in free space are to be determined, the exterior boundary S_2 is allowed to recede to infinity and in this case the boundary conditions on S_2 are the radiation conditions :

$$\lim_{R\to\infty} \begin{cases} R\underset{\sim}{E}(\underset{\sim}{R}) \; , \; R\underset{\sim}{H}(\underset{\sim}{R}) \quad \text{bounded} \\ R[\underset{\sim}{E}(\underset{\sim}{R}) + (\mu/\epsilon)^{\frac{1}{2}}\hat{\underset{\sim}{R}} \times \underset{\sim}{H}(\underset{\sim}{R})] = 0 \\ R[\underset{\sim}{H}(\underset{\sim}{R}) - (\epsilon/\mu)^{\frac{1}{2}}\hat{\underset{\sim}{R}} \times \underset{\sim}{E}(\underset{\sim}{R})] = 0 \end{cases} \tag{2.7}$$

where $\underset{\sim}{R} = R\hat{\underset{\sim}{R}}$ is a vector to some point on S_2. These conditions ensure that the solution fields will represent outward going radiation with $\underset{\sim}{E}$ transverse to $\underset{\sim}{H}$ and both transverse to the direction of propagation at S_2.

Equations (2.3) - (2.5) and the boundary conditions (2.6), (2.7) are sufficient to define a unique solution for arbitrary S_1.

An application of the Green's function technique [3] to this boundary value problem gives the "electric field integral equation" (EFIE) and "magnetic field integral equation" (MFIE) :

$$\underset{\sim}{E}(\underset{\sim}{r}) = \tau\underset{\sim}{E}^{inc}(\underset{\sim}{r}) + \tau \int_{S_1} i\omega\mu\underset{\approx}{G}_0(\underset{\sim}{r},\underset{\sim}{r}') \cdot \underset{\sim}{J}_s(\underset{\sim}{r}')dS_1' \quad \text{(EFIE)} \tag{2.8}$$

$$\underset{\sim}{H}(\underset{\sim}{r}) = \tau\underset{\sim}{H}^{inc}(\underset{\sim}{r}) + \tau \int_{S_1} \nabla' \times \underset{\approx}{G}_0(\underset{\sim}{r},\underset{\sim}{r}') \cdot J_s(\underset{\sim}{r}')dS_1' \quad \text{(MFIE)} \tag{2.9}$$

Here, $\underset{\approx}{G}_0$ is the free space dyadic Green's function for the vector wave equation, $\underset{\sim}{J}_s(\underset{\sim}{r}') = \hat{\underset{\sim}{n}} \times \underset{\sim}{H}(\underset{\sim}{r}')$ is the surface current density on S_1 and $\underset{\sim}{E}^{inc}$, $\underset{\sim}{H}^{inc}$ are the incident fields. The integrals are to be interpreted as principal values and the prime on the operator ∇ indicates that it acts only on $\underset{\sim}{r}'$. The constant factor τ is required because of the singularity of the Green's function at $\underset{\sim}{r} = \underset{\sim}{r}'$, and has the value

$$\tau = \begin{cases} 1 & \underset{\sim}{r} \text{ not on } S_1 \\ 2 & \underset{\sim}{r} = \underset{\sim}{r}' \text{ on a smooth } S_1 . \end{cases}$$

At a discontinuity on S_1, τ has a value related to the subtended solid angle at the discontinuity [2].

The unknown surface current distribution $\underset{\sim}{J}_s$ on a smooth S_1 is found by applying the boundary condition (2.6) in the EFIE and MFIE :

$$\hat{\underset{\sim}{n}} \times \underset{\sim}{E}^{inc}(\underset{\sim}{r}') = -\hat{\underset{\sim}{n}} \times \int_{S_1} i\omega\mu \underset{\approx}{G}_0(\underset{\sim}{r},\underset{\sim}{r}') \cdot \underset{\sim}{J}_s(\underset{\sim}{r}')dS_1' \qquad (2.10)$$

$$\underset{\sim}{J}_s(\underset{\sim}{r}') = 2\hat{\underset{\sim}{n}} \times \underset{\sim}{H}^{inc}(\underset{\sim}{r}') + 2\int_{S_1} \nabla' \times \underset{\approx}{G}_0(\underset{\sim}{r},\underset{\sim}{r}') \cdot \underset{\sim}{J}_s(\underset{\sim}{r}')dS_1' \qquad (2.11)$$

which gives first and second kind linear Fredholm equations for $\underset{\sim}{J}_s$ in terms of the known incident fields. The field at any other point in the medium can then be found by substituting this current distribution into the EFIE or MFIE.

In physical terms, the incident field induces a current distribution on the surface of the reflector so that the total field at the surface satisfies the boundary condition (2.6). The total field in the medium is then the sum of the incident field and the field radiated by this current distribution.

Closed form solutions of these equations are known only for a few simple surfaces such as the sphere, cone and spheroid, and in general scattering solutions are found either numerically, by a reduction of the integral equations (2.10), (2.11) to matrix form [4], [5], [6] or by the use of some analytical approximations to them.

A difficulty which can arise in the numerical solution of (2.10) or (2.11) is that the homogeneous equations (i.e. those with no incident fields) have non-trivial solutions at certain discrete values of k corresponding to interior electric or magnetic modes of oscillation in S_1 . The consequent non-uniqueness results in severe numerical instability over a range of values of k near these eigenvalues and corrective measures must be taken. These methods are well documented [7] and involve the use of a combination of the EFIE and MFIE having a unique solution for all k , or an extension of the boundary conditions by setting the field to zero at a discrete set of points inside S_1 [8] .

It is interesting to note that regularisation and related stabilisation techniques are not appropriate for this "non-uniqueness instability" as they result in the complete suppression of the resonant current distribution (corresponding to the particular eigenvalue responsible for the instability), which may in fact form part of the correct solution of the inhomogeneous equations.

For scattering bodies with dimensions of the order of a wavelength, a numerical solution of the integral equations by matrix methods is usually necessary, and economical. But the computational effort required for scatterers larger than a few wavelengths becomes excessive and although moment method calculations for large reflectors have been reported [18,19], approximations are usually made to simplify or even eliminate the integral equations.

Only two approximations useful for reflector antennas will be mentioned here and the reader is referred to the literature for further information on this extensive subject. The geometric optics approximation effectively assumes that the propagating wave has zero wavelength and as a consequence that scattering can be adequately described by the ray laws of optics. Since diffraction effects are completely neglected, this approximation is most accurate for large reflectors and for observing positions which receive a specular reflection.

A better approximation to the scattered field is given by physical optics, which approximates the surface current distribution at each illuminated point on the scatterer by that which would appear on a tangent plane at the point if the incident radiation were a plane wave. On the shadowed part of the scatterer, the current is assumed to be zero. That is,

$$\underset{\sim}{J}_s(\underset{\sim}{r}') = \begin{cases} 2\hat{\underset{\sim}{n}} \times \underset{\sim}{H}^{inc}(\underset{\sim}{r}') & \text{on the illuminated part of } S_1 \\ 0 & \text{on the shadowed region .} \end{cases} \tag{2.12}$$

Equation (2.11) shows that this is equivalent, on the illuminated part of S_1 , to neglecting the integral term which represents the mutual interaction of the surface currents. The scattered field is ultimately calculated by using (2.12) in the EFIE or MFIE and thus satisfies Maxwell's equations. Although the approximation is in principle restricted to large scatterers with large radii of curvature, it successfully predicts the forward-scattered field even for relatively small smooth reflectors, and has been widely applied in the analysis of microwave scattering problems.

3. Reflector Synthesis

In terms of the scattering problem, reflector synthesis requires the determination of the surface S_1 which will generate a specified scattered field (a "shaped beam") when illuminated by a known incident field. The direct use of the EFIE or MFIE in this application is difficult because of the two-stage nature of the scattering calculation, requiring an intermediate determination of $\underset{\sim}{J}_s$ on the as yet unknown surface. However, a useful formulation can be found if one of the approximations mentioned above can be applied.

The geometric optics approximation leads to a well-known synthesis method [9] giving the reflector shape in closed form. But the method is unsuitable for the design of electrically small reflectors where the neglected diffraction effects can form a significant part of the scattered field. The physical optics approximation correctly represents these effects but at the cost of no longer allowing an analytical solution of the synthesis problem. The practical advantages of small reflectors have nevertheless stimulated interest in synthesis methods using this approximation.

Substitution of the physical optics approximation for the surface current (eqn 2.12) in the EFIE and MFIE gives the following expressions for the scattered field :

$$\underset{\sim}{E}^{scat}(\underset{\sim}{r}) = 2i\omega\mu \int_{S_I} \underset{\approx}{G}_0(\underset{\sim}{r},\underset{\sim}{r}') \cdot \hat{\underset{\sim}{n}} \times \underset{\sim}{H}^{inc}(\underset{\sim}{r}')dS_I' \tag{3.1}$$

and

$$\underset{\sim}{H}^{scat}(\underset{\sim}{r}) = 2 \int_{S_I} \nabla' \times \underset{\approx}{G}_0(\underset{\sim}{r},\underset{\sim}{r}') \cdot \hat{\underset{\sim}{n}} \times \underset{\sim}{H}^{inc}(\underset{\sim}{r}')dS_I' \tag{3.2}$$

where S_I is the illuminated part of the scatterer, usually the front surface of the reflector. The first equation can be rewritten in operator notation as

$$E = K\rho \tag{3.3}$$

where E is the scattered electric field, ρ is a function describing the reflector surface and K is an operator defined by (3.1). The essential nonlinearity of the problem arises in this crucial step. In terms of the synthesis problem, eqn (3.1) expresses the scattered field as an undetermined linear operator (integration over the unknown S_I) acting on a known function, while eqn (3.3) expresses the field as a known operator, now generally non-linear for geometric reasons, acting on the unknown function ρ representing the surface.

In many applications, the phase of the scattered field is unimportant and only the radiated power P (or "power pattern") is specified. Since $P = \beta E^*E$ where β is a constant of the medium and the $*$ indicates a complex conjugate, eqn (3.3) can be modified to

$$P = T\rho$$

where

$$T = \beta K^*K \,. \tag{3.4}$$

In order to determine the reflector shape, this equation must be solved for ρ when P is given. The immediate difficulty is that in general no solution exists, since an arbitrary P will not lie in the range of T when the function ρ is bounded. For example, it can be shown that a discontinuous field, which is often desirable in practice, cannot be generated by a finite reflector [17].

However it is usually sufficient in practical applications to require that the radiation pattern of the synthesised reflector be close to the desired pattern $\hat{P}$ in the sense of satisfying an upper bound on the approximation error $\|T\rho-\hat{P}\|$ for some norm $\|\cdot\|$. This leads to a restatement of the synthesis problem as : find $\hat{\rho}$ such that

$$\|T\hat{\rho}-\hat{P}\| = \min \|T\rho-\hat{P}\| \quad \text{for} \quad \rho \in D(\rho) \tag{3.5}$$

where $D(\rho)$ is the set of allowable ρ .

Local minima of $\|T\rho-\hat{P}\|$ can be found numerically by using an iterative procedure based on a linearisation of T to improve an initial estimate of $\hat{\rho}$. Gradient search or variational methods using the Gâteaux derivative of the error functional $\|T\rho-\hat{P}\|$ have been successfully applied in a wide range of synthesis problems [10,11]. An alternative method is possible when the Fréchet derivative L_ρ of T, i.e. the continuous linear operator L_ρ such that

$$\lim_{\delta\rho\to 0} \frac{1}{\|\delta\rho\|} (T(\rho+\delta\rho) - T\rho - L_\rho\delta\rho) = 0 \text{ ,}$$

can be found or approximated.

If $\rho = \rho_0 + \delta\rho_0$, where ρ_0 is an initial estimate of $\hat{\rho}$, then

$$T\rho - \hat{P} = T(\rho_0+\delta\rho_0) - \hat{P}$$

$$= L_\rho\delta\rho_0 - Q_0 \qquad \text{for small } \delta\rho_0$$

where $Q_0 = \hat{P} - T\rho_0$ can be identified as the approximation error in using ρ_0 as the reflector surface. The solution of the linearised problem : find $\delta\rho_n$ such that

$$\|L_\rho\delta\rho_n - Q_n\| = \min \|L_\rho\delta\rho - Q_n\| \text{ , } \quad \rho_n + \delta\rho \in D(\rho)$$

is the Newton correction [12] and the sequence

$$\rho_{n+1} = \rho_n + \delta\rho_n \qquad n = 0,1,2,\ldots$$

converges to a local minimum if the initial estimate ρ_0 is close enough to it.

4. Numerical Example

An application of this method occurs in the design of the azimuth antenna for the Interscan microwave landing system [13]. This antenna radiates a fan beam which is electronically scanned from side to side to enable an aircraft to determine its azimuth angle relative to the runway. In a horizontal plane, the beam is narrow in order to provide high angular resolution, while in elevation the beam is required to approximate the ideal pattern in Figure 2, where an additional "sidelobe constraint" $P < P_{SL}$ for $\theta < \theta_{SL}$ is specified to minimise ground reflection effects at low observation angles.

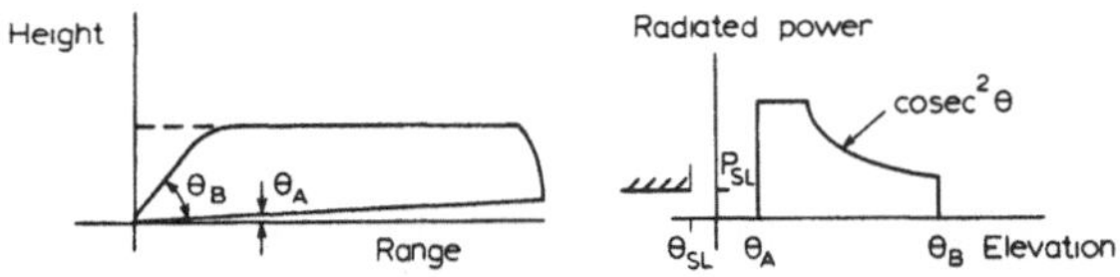

Figure 2. Ideal elevation patterns (a) in space and (b) as a function of elevation angle θ.

An antenna which has been developed to generate a fan beam of this kind is shown in Figure 3, and consists of a cylindrical reflector illuminated by a directly radiating parallel plate lens [14]. The elevation pattern of this configuration is determined by the cross-sectional profile of the reflector. For observation points distant from the antenna, the radiated field given by eqn. (3.1) for a transversely polarised incident field can be separated into

$$\underset{\sim}{E}^{scat}(\underset{\sim}{r}) = \hat{\underset{\sim}{1}}_\theta f(\underset{\sim}{r})E_\theta(\theta) \cdot E_\gamma(\gamma)$$

where θ and γ are the elevation and azimuth angles and $\hat{\underset{\sim}{1}}_\theta$ is a unit vector in the direction of increasing θ. The elevation pattern E_θ is given by

$$E_\theta(\theta) = A\int_{\varphi_1}^{\varphi_2} S(\varphi)\cos\left[\varphi+\theta-\tan^{-1}\left(\frac{\dot{\rho}}{\rho}\right)\right] \cdot \left[\rho + \frac{\dot{\rho}^2}{\rho}\right]^{\frac{1}{2}} e^{-ik\rho(1+\cos(\varphi+\theta))}d\varphi$$

where $\rho(\varphi)$ is the reflector profile, A is almost independent of ρ, $S(\varphi)$ is the intensity of the incident field at the reflector and the other quantities are defined in Figure 4. Dot notation is used to signify differentiation with respect to φ.
This equation can be written as

$$E_\theta(\theta) = \int_{\varphi_1}^{\varphi_2} \alpha(\rho,\dot{\rho},\theta,\varphi)e^{-i\Phi(\rho,\theta,\varphi)}d\varphi$$

corresponding to equation (3.3). K is thus a nonlinear first kind Fredholm operator. A suitable form for L_ρ can be found by noting, after Daveau [16], that the main effect of a change $\delta\rho$ in ρ occurs through the phase term $\Phi(\rho,\theta,\varphi)$. Neglecting the dependence of α on ρ,

$$L_\rho\delta\rho = \int_{\varphi_1}^{\varphi_2} h(\rho,\theta,\varphi)\delta\rho(\varphi)d\varphi$$

where

$$h = 2\mathrm{Re}\left(-i\alpha E^* e^{i\Phi}\frac{\partial\Phi}{\partial\rho}\right).$$

Thus the Newton correction $\delta\rho_n$ is the solution (in the sense of 3.7) of

$$\int_{\varphi_1}^{\varphi_2} h(\rho_n,\theta,\varphi)\delta\rho_n d\phi = Q_n(\theta)$$

or

$$L_\rho\delta\rho_n = Q_n. \tag{4.1}$$

This is clearly a linear first kind of Fredholm equation and can be reduced to a system of linear algebraic equations by a number of well-known methods. A general collocation method [15] is used here. Briefly, $\delta\rho$ is expressed as a linear combination (with coefficients $\underset{\sim}{a}$) of N triangle functions $\{t_j\}$ on $[\varphi_1,\varphi_2]$ and collocation is applied at M points $\theta_1 < \theta_2 < \ldots < \theta_M$. Thus (4.1) becomes

$$[\ell_{ij}]\underset{\sim}{a} = \underset{\sim}{q}$$

where $[\ell_{ij}]$ is the matrix representation of L_ρ with

$$\ell_{ij} = \int_{\varphi_1}^{\varphi_2} h(\rho_n,\theta_i,\varphi)\,t_j(\varphi)\,d\varphi \qquad i = 1,\ldots,M\ ;\quad j = 1,\ldots,N$$

and

$$\underset{\sim}{q} = \begin{bmatrix} Q_n(\theta_1) \\ \vdots \\ Q_n(\theta_M) \end{bmatrix} .$$

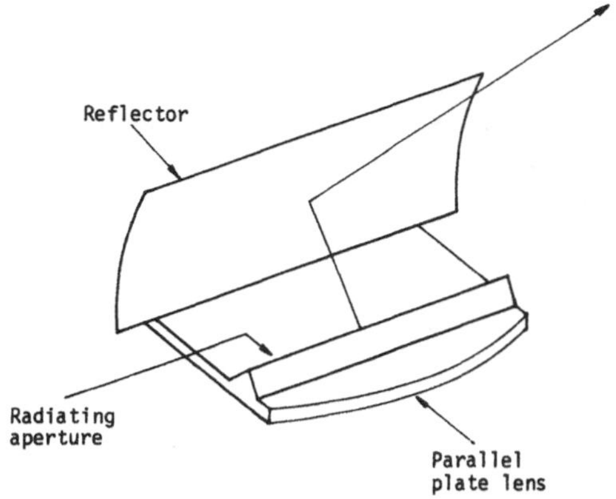

Fig. 3 - Lens-fed reflector antenna.

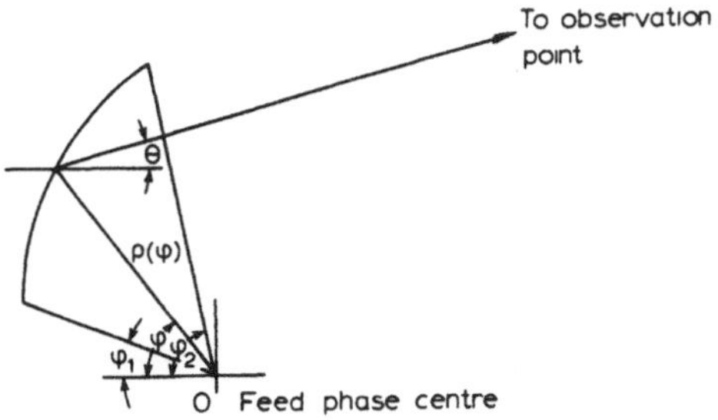

Figure 4. Geometry of the reflector profile.

The class of allowed $\delta\rho$ is restricted to those for which $\delta\rho(\varphi_1) = 0$ by omitting the first (single sided) triangle function in the usual complete basis on $[\varphi_1, \varphi_2]$. This ensures that the bottom of the reflector remains fixed.

A solution $\underset{\sim}{a}$ is sought which minimises

$$\|\underset{\sim}{w}^T([\ell_{ij}]\underset{\sim}{a}-\underset{\sim}{g})\|^2 + \eta\|\underset{\sim}{a}\|^2 \tag{4.2}$$

where a weight vector $\underset{\sim}{w}$ is used to allow further control of the approximation error over the range $[\theta_1, \theta_M]$ and zeroth order regularisation is used to stabilise the system. Writing $W = \mathrm{diag}(\underset{\sim}{w})$, the solution of (4.2) is

$$\underset{\sim}{a} = ([\ell_{ij}]^T W^2 [\ell_{ij}] + \eta I)^{-1} [\ell_{ij}]^T W^2 \underset{\sim}{g}$$

giving the Newton correction $\delta\rho = \underset{\sim}{a}.\underset{\sim}{t}$.

Figure 5 shows the result of this procedure when a geometric optics design is used as the initial approximation. The weight vector in this case was adjusted to reduce the error for $\theta < 0°$ at the expense of the region $\theta > 15°$ in order to satisfy the below-horizon sidelobe constraint. Some indication of the instability of the problem is given by the similarity of the initial and optimised profiles which differ by a maximum of 0.15 wavelengths.

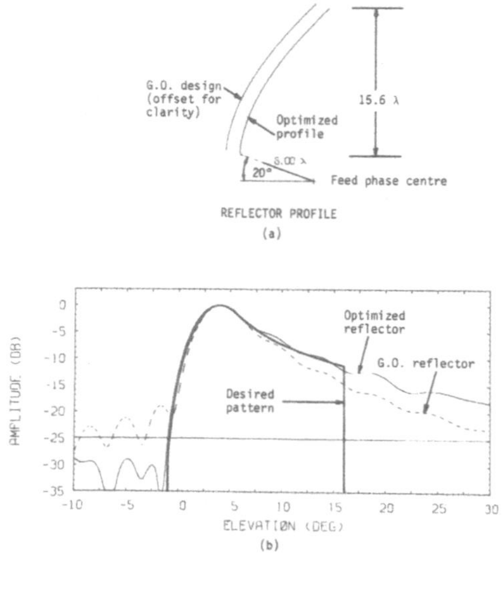

Figure 5. (a) Geometric optics and optimized reflector profiles.

(b) Radiation patterns.

5. Conclusion

The synthesis method described here, and others based on linearisation, cannot guarantee a globally optimum solution, or even a locally optimum solution unless a good initial approximation is known. Nevertheless, these techniques have been widely applied and usually provide a significant improvement in the reflector performance over geometric optics designs. The Newton iteration, when it can be applied, offers a substantial reduction in computer time and provides a convenient framework in which engineering design criteria can be applied.

Acknowledgements

The author gratefully acknowledges the assistance of Dr G.T. Poulton and members of the Applied Research Group in the Division of Radiophysics. This work was partially supported by Department of Productivity funding of the Interscan project.

References

[1] de Hoop, A.T., 1977. "General considerations on the integral-equation formulation of diffraction problems", in *Modern Topics in Electromagnetics and Antennas*, Peter Peregrinus.

[2] Poggio, A.J. and Miller, E.K., 1973. "Integral equation solutions of three-dimension scattering problems", in *Computer Techniques for Electromagnetics*, Mittra, R. (Ed), Pergamon Press.

[3] Tai, C.-T., 1971. *Dyadic Greens Functions in Electromagnetic Theory*, Intext Educational Publishers.

[4] Harrington, R.F., 1968. *Field Computation by Moment Methods*, Macmillan.

[5] Davies, J.B., 1977. "Numerical approaches to electromagnetic problems", in *Modern Topics in Electromagnetics and Antennas* op. cit.

[6] Jones, D.S., 1974. "Numerical methods for antenna problems". *Proc. IEE* 121, 573.

[7] Mittra, R. and Klein, C.A., 1975. "Stability and convergence of moment method solutions", in *Numerical and Asymptotic Techniques in Electromagnetism*, Mittra, R. (Ed.), Springer.

[8] Schenck, H.A., 1968. "Improved integral formulation for acoustic radiation problems" *J. Acoust. Soc. Am.* 44, 41.

[9] Silver, S. (ed.), 1965. *Microwave Antenna Theory and Design*, Dover, p.497.

[10] Clarricoats, P.J.B., 1979. "Some recent advances in microwave reflector antennas". *Proc. IEEE* 126, 9.

[11] Clarricoats, P.J.B. and Poulton, G.T., 1977. "High efficiency microwave reflector antennas - a review". *Proc. IEEE* 65, 1470.

[12] Rall, L.B., 1969. *Computational Solution of Nonlinear Operator Equations*, Wiley.

[13] Minnett, H.C., 1975. "The Interscan system". *IREE Conv. Digest*, p.238.

[14] Poulton, G.T., 1979. "Directly radiating lens antennas for Interscan". To be published in *IREE (Aust.) Conv. Digest*.

[15] Hanson, R.J. and Phillips, J.L., 1975. "An adaptive numerical method of solving linear Fredholm integral equations of the first kind". *Numer. Math.* 24, 291.

[16] Daveau, B., 1970. "Synthese et optimisation de réflecteurs de forme spéciale pour antennes". *Rev. Tech. Thomson-CSF* 2, 37.

[17] Deschamps, G.A. and Cabayan, H.S., 1972. "Antenna synthesis and solution of inverse problems by regularisation methods". *IEEE Trans. Antennas Propag.* AP-20, 268.

[18] Rusch, W.V.T., 1975. "Reflector Antennas", in *Numerical and Asymptotic Techniques in Electromagnetism* op. cit.

[19] Kinzel, J.A., 1974. "Large reflector antenna pattern computation using moment methods". *IEE Trans. Antennas Propag.* AP-22, 116.

SYSTEMS IDENTIFICATION AND ESTIMATION FOR CONVOLUTION INTEGRAL EQUATIONS

A.J. Jakeman and P. Young
Centre for Resource and Environmental Studies, Australian National University, Canberra, A.C.T.

1. Introduction

In this paper, we consider the solution of the Volterra convolution equation of the first kind

$$x(t) = \int_0^t k(t-\tau)u(\tau)d\tau \ . \qquad (1.1)$$

Equation (1.1) often arises in the analysis of systems where an input $u(t)$ excites a system with impulse response $k(t)$ to yield, by convolution, an output $x(t)$. The applications of this equation are many and may be divided into two broad categories: those for which the unknown function is

(a) a fixed and regular impulse response function of the type met in the solution of linear ordinary differential equations; and
(b) of continually changing form with no obvious underlying finite parametrization.

The former corresponds to solving (1.1) for $k(t)$ given $u(t)$ and $x(t)$ while the latter corresponds to solving (1.1) for $u(t)$ given $k(t)$ and $x(t)$.

In order to demonstrate how the two cases occur in practice, we make use of a simple hydrological example. We see that case (a) concerns the estimation of the kernel $k(t)$ from measurements of input rainfall data $u(t)$ and output run-off flow data $x(t)$. Here $k(t)$ is referred to as the impulse response or the 'input hydrograph'. When (1.1) is equivalent to a linear differential equation with constant coefficients, then the Laplace transform of the kernel $k(t)$ (known as the transfer function of $k(t)$ in the engineering and control literature) is a rational function, the parametrization of which is finite. For case (a), the strategy is to approximate the transfer function by such a rational function. In case (b), on the other hand, the analyst is presented with measurements of $x(t)$ and an estimate of $k(t)$ and is required to estimate the input rainfall $u(t)$. As we shall see, this is a much more difficult problem whose solution may be dependent upon factors such as the level of noise on the measurement of $x(t)$ and the 'invertibility' of the system characterised by the impulse response $k(t)$.

Many systems analysis problems can be considered as examples of case (a) and the simple transfer function approach has proved very successful in practical applications. A number of these applications are discussed in the paper and representative results are presented from the analysis of data from chemical (fluorescence decay), medical (transport properties of the human circulatory system), botanical (transport properties of problem flow in plants), hydrological, engineering and socio-economic systems.

2. Impulse Response Estimation

In order to show how generally the problem of impulse response estimation can be formulated, we begin with the following multivariable form of equation (1.1),

$$\underline{x}(t) = \int_0^t \underline{K}(t-s)\underline{u}(s)ds \ . \tag{2.1}$$

Here, $\underline{x}$ and $\underline{u}$ are vectors of p outputs and q inputs, respectively, and $\underline{K}$ is a matrix of order $p \times q$. For example, the ℓ-th (output) equation is

$$x_{(\ell)}(t) = \int_0^t \sum_{j=1}^{q} k_{(\ell j)}(t-s)u_{(j)}(s)ds \ , \qquad \ell = 1(1)p \ . \tag{2.2}$$

Let $f(s) = L\{f(t)\}$ denote the Laplace transform of a continuous function $f(t)$ and D denote the derivative operator. Then (2.1) can be transformed to

$$\underline{x}(s) = L\{\underline{K}(t)\}\underline{u}(s) \tag{2.3}$$

where $\underline{x}(s)$ and $\underline{u}(s)$ are, respectively, the Laplace transforms of $\underline{x}(t)$ and $\underline{u}(t)$.

For linear systems, (2.3) can be replaced by

$$\underline{A}(s)\underline{x}(s) = \underline{B}(s)\underline{u}(s) \tag{2.4}$$

where

$$\underline{A}(s) = I_p + A_1 s + A_2 s^2 + \dots + A_r s^r \tag{2.5}$$

and

$$\underline{B}(s) = B_0 + B_1 s + B_2 s^2 + \ldots + B_{r-1} s^{r-1} \ . \tag{2.6}$$

All the matrices A are square of order p while all the matrices B are $p \times q$, and I_p is the identity of order p. Note that some of the B matrices may be zero and particularly that r is finite and usually of small size. For the scalar case with single input and output, (2.4) becomes

$$\frac{b_0 + b_1 s + \ldots + b_{r-1} s^{r-1}}{1 + a_1 s + \ldots + a_r s^r} \tag{2.7}$$

which defines the rational "transfer function" of k(t) (viz. L{k(t)}). Thus, the TF model (2.4) is a parametrically efficient representation of the more general situation defined by (2.1). By inverting (2.4) back to the time domain we obtain the continuous-time ordinary differential equation model

$$\underline{A}(D)\underline{x}(t) = \underline{B}(D)\underline{u}(t) \ . \tag{2.8}$$

Often, in practice, it is necessary to work with discrete-time models. It can be shown [1] that a differential equation model like (2.8) can be transformed into a discrete-time TF model.

$$\underline{A}(z^{-1})\underline{x}_i = \underline{B}(z^{-1})\underline{u}_i \tag{2.9a}$$

where $\underline{x}_i$ and $\underline{u}_i$ are the values of $\underline{x}(t)$ and $\underline{u}(t)$ at an arbitrary i-th sampling instant and $z^{-\ell}$ is the backward shift operator so that $z^{-\ell}\underline{x}_i = \underline{x}_{i-\ell}$. The z nomenclature is used here because, in simple terms, z can be identified with the z transform operator. The $\underline{A}(z^{-1})$ and $\underline{B}(z^{-1})$ are matrix polynomials in z^{-1} of the following general form

$$\underline{A}(z^{-1}) = I + A_1 z^{-1} + \ldots + A_n z^{-n}$$

$$\underline{B}(z^{-1}) = B_0 + B_1 z^{-1} + \ldots + B_m z^{-m} \ .$$

The dynamic orders r of the continuous-time model and n of the discrete-time model are not necessarily the same.

In practice, it is necessary to allow for the stochastic nature of most systems and to incorporate errors into the model formulation. In the discrete-time case, for example, this can be accomplished quite simply by assuming that the measured output is denoted by $\underline{y}_i$ where

$$\underline{y}_i = \underline{x}_i + \underline{\xi}_i \tag{2.9b}$$

and $\underline{\xi}_i$ is a vector of errors. Whilst this specific method of introducing stochastic elements into the models nominally restricts the class of stochastic systems considered, it appears that many practical systems can be represented in this manner. For the proposed estimation procedure, the only necessary assumption on the errors $\underline{\xi}_i$ is that they are uncorrelated with the input excitation $\underline{u}_i$.

Equations (2.9a) and (2.9b) provide a parametrically efficient approximation to the convolution integral equation (2.1). There are two stages for obtaining the unknown parameters in such an approximation from observational data $\underline{u}_i, \underline{y}_i$ $i = 1,2,\ldots,N$. First, the structure of the models, like (2.8) or (2.9a), must be *identified.* In the scalar discrete-time case, this entails estimation of the value of n and m . Second, the parameters characterising the structure estimated in the first stage must themselves be *estimated* from the observational data. For both of these stages, we employ an instrumental variable (IV) technique (for details, see Appendix). This is a simple modification of the well-known least squares procedure, which avoids the inherent limitations of least squares analysis when applied to dynamic system estimation problems [2]. It can be shown that the IV method yields estimates which are consistent and relatively efficient in statistical terms. If, in addition, $\underline{\xi}_i$ can be assumed to be generated from white noise by a multivariable (or vector) autoregressive moving average process, then a "refined" IV technique can be used, which yields asymptotically efficient parameter estimates. Details of these procedures, which are usually implemented in recursive form, are well documented and can be found in [3-8].

It should be noted that several alternative procedures can be used. Usually, it is sufficient and convenient to employ the basic IV method for the order identification stage. However, it is useful to invoke refined IV for the parameter estimation stage since this algorithm provides a good estimate of the variance-covariance matrix of the parametric errors.

These two stages of identification and estimation yield a rational transfer function approximation of $k(t)$. If the impulse response itself is required then it can be evaluated quite easily. In the discrete-time scalar case, for example, k_i is the ith term in the nominally infinite series obtained on dividing $B(z^{-1})$ by $A(z^{-1})$. Alternatively, the k_i can be obtained by solving (2.9a) recursively with u_i defined as a discrete-time impulse of magnitude $1/T$ at the origin. If T is the uniform width of the data sampling interval, then $u_i = \delta_{i,1}/T$.

2.1 A simple Hydrological Example

To demonstrate the application of this TF procedure, consider the hydrological problem where a rainfall measure is assumed to be the input u_i $(i = 1,2,\ldots,N)$ to a catchment system with impulse response or "unit hydrograph" k_i $(i = 1,2,\ldots,\infty)$ and output y_i $(i = 1,2,\ldots,N)$ which is the run-off or 'observed hydrograph'. In fact, this rainfall measure is the 'effective' rainfall rather than the actual rainfall. It is obtained by accounting for soil moisture and evaporation effects in the particular catchment. This is mentioned again in section 4.4 but for a detailed explanation the reader is referred to [9].

In some hydrological applications, the data k_i and y_i are supplied and the u_i must be determined. This, of course, corresponds to input estimation, which will be discussed in the next section. As will be shown there, this more difficult problem can only be analysed after a simple parametrization for the k_i data has been constructed. Since the k_i represent the output of a system that has been excited by a unit impulse, the discrete transfer function modelling described above can be applied. The IV identification and estimation procedure yielded the following difference equation model with $n = 2$ and $m = 3$:

$$\frac{.03 + .0461z^{-1} + .01488z^{-2}}{1.14635z^{-1} + .55331z^{-2}} \; . \tag{2.10}$$

The fit of this model to the k_i data (circled) is displayed in Figure 1 with the residuals plotted below the time axis. It is clear that the transfer function model is a very good description of the system impulse response and so provides a parametrically efficient (only 5 parameters) representation of the system. In the next section, we will pursue this example further in order to try to estimate the rainfall input u_i .

We note in passing that any decay function data can be approximated by a rational transfer function of the form (2.7) by applying the IV identification and estimation techniques to input and output data consisting of a unit impulse and the decay data, respectively.

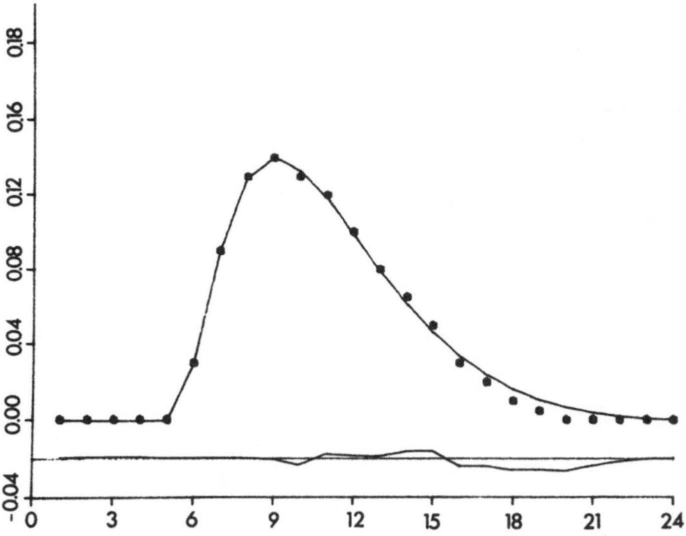

Figure 1 : The fit of the TF model (2.10) to the unit hydrograph data. Residuals are plotted below.

3. Input Estimation

As we have pointed out, the problem of input estimation (the true "inverse problem") is not nearly so straightforward as the transfer function or impulse response estimation problem. To illustrate this, let us first consider the simple hydrological problem and then progress to the more general situation.

3.1 The Simple Hydrological Example

In this example, the unit hydrograph and observed hydrograph data to be used for the estimation of effective rainfall input are provided in a smoothed form. In other words, there is little apparent noise on these data as can be seen from Figures 1 and 5. Since y_i is approximately equivalent to x_i therefore, the simplest approach to the estimation of u_i is to try and solve the equation

$$u_i = \frac{A(z^{-1})}{B(z^{-1})} y_i \tag{3.1}$$

for u_i, knowing y_i and the transfer function as estimated in (2.10). In other words

$$.03u_i = y_i - 1.4635y_{i-1} + .55331y_{i-2} - .04188u_{i-2} \,. \tag{3.2}$$

The result of performing this TF inversion is shown in Figure 2 which is clearly unacceptable The reason for such an unstable result is that the transfer function in (2.10) is non-invertible. Invertibility of a transfer function requires that the roots of its numerator polynomial $B(z^{-1})$, in terms of z^{-1}, lie outside the unit circle. Since it is clear in this case that one of the roots of $B(z^{-1})$ lies inside the unit circle, it follows that the inverse operation in (3.1) and (3.2) is unstable, as indicated in Figure 2. It should be noted that this concept of non-invertibility of an operator in time-series analysis is closely related to that of numerical instability for ordinary differential equations as discussed in [18].

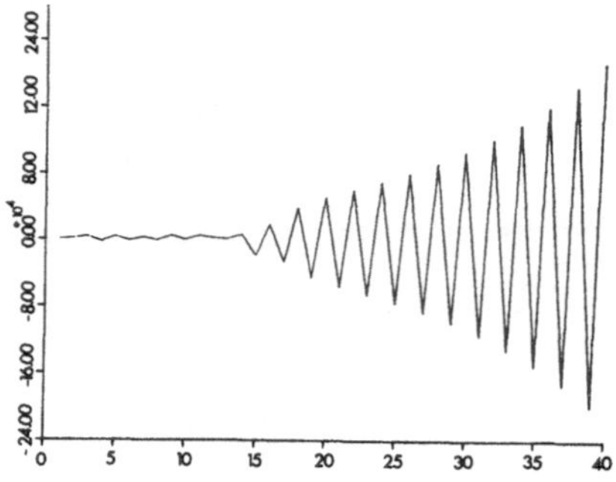

Figure 2 : The unstable input rainfall generated from (3.2) which involves the inversion of (2.10).

Because the dynamics of a linear process described by a rational TF are dominated by the $\underline{A}$ (or $\underline{A}$) polynomial and the $\underline{B}(z^{-1})$ merely describes the initial transient portion of the impulse response, it is usually easy to find a dynamic equivalent of a rational TF which is invertible. In this case, one invertible approximation is as follows

$$\frac{B(z^{-1})}{A(z^{-1})} = \frac{.09098z^{-1}}{1-1.4635z^{-1} + .55331z^{-2}} \quad . \tag{3.3}$$

This is not the only approximation, of course, but will suffice for the present illustrative purposes. In fact, for this particular approximation, the following properties are equivalent to those of the model (2.10) :

(i) the steady state gain, $B(1)/A(1)$, which is the area under the impulse response

(ii) the time constants which are obtained from the zeros α_i of $A(z)$ as $\tau_i = -1/\log(\alpha_i)$ when $A(z)$ has all real roots (see [1])

(iii) the instant of decay of the impulse response

(iv) the peak or amplitude of the impulse response

The relationship of the impulse response of this approximate TF to the 'true' unit hydrograph is shown in Figure 3. Solving (3.1) now with $B(z^{-1})$ redefined as in (3.3) yields the input plotted in Figure 4. This too seems physically unacceptable, since it indicates negative estimated rainfall over a short period between $k = 14$ and 16; it is, however, a stable result. Of course, the operation of (3.3) on this estimated input yields a perfect fit to the output data. Interestingly, it was also found that the operation of (2.10) on this same input produced a quite reasonable fit to the output data as can be seen from Figure 5. Hence this very simple method seems to work provided the model (1.1) is appropriate and the k_i and y_i data are accurate and relatively noise-free. Clearly, in this example, either the model (1.1) is inappropriate or the data are not good representations.

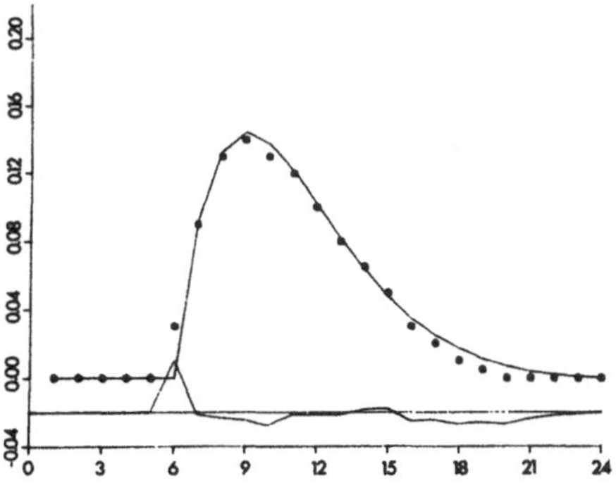

Figure 3 : The fit of the TF model (3.3) to the hydrograph. This is an invertible dynamic equivalent of (2.10).

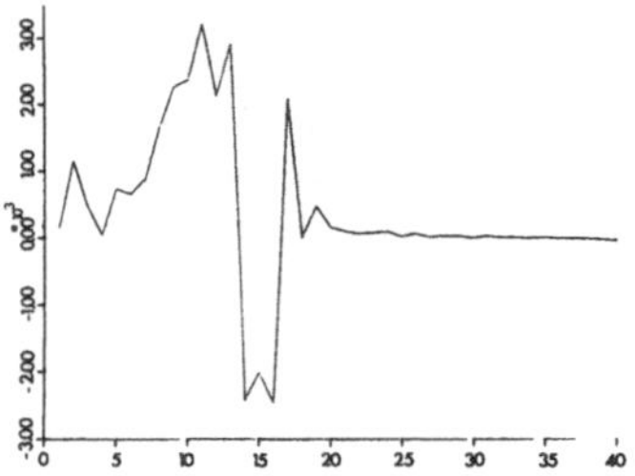

Figure 4 : The stable input rainfall generated from inversion of (3.3).

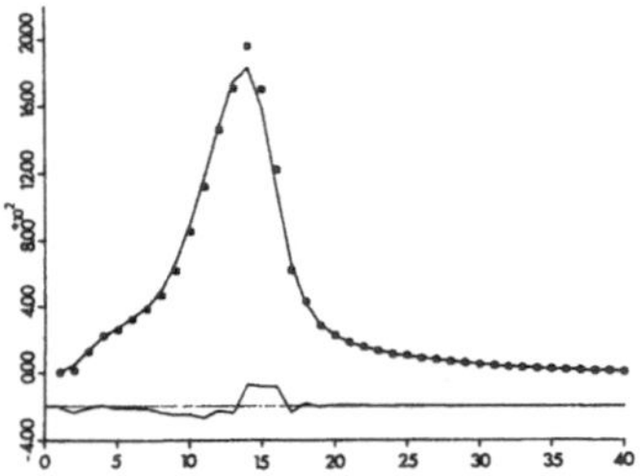

Figure 5 : The fit of the TF model (2.10) to the output data using the input displayed in Figure 4.

3.2 The General Problem of Input Estimation

It is clear from the above example that the problem of input estimation is concerned with the estimation of a time-varying signal in the presence of noise. In control and systems terms this is referred to as a 'filtering' problem and much work has been written on this subject, particularly in the last forty years. A general solution to the linear filtering problem was suggested by Kalman [10] and the recursive Kalman filter, as it is called, has been used increasingly to solve problems in which the 'state variables' of a dynamic system need to be estimated from noisy measurements.

The input estimation problem itself is not necessarily a straightforward filtering problem since it is the input and not the state variables (coefficients in the rational transfer function in the present application) that have to be estimated. This has not been considered seriously in the control and systems literature, although a recursive approach similar to that of Kalman is possible. This is discussed more fully by Young and Jakeman [11]. Their approach entails the recursive estimation of the input variable under the assumption that it can be described by a Gauss-Markov stochastic model. Initial simulation results using this approach have yielded encouraging results [11]. It is also interesting to note that the recursive Kalman filter is similar in form to the recursive IV algorithm discussed here. Indeed, this similarity has allowed the IV method to be extended to the estimation of time-variable parameters [7,12].

In some contexts, there exist other approaches to the problem of input-type estimation. We use the terminology input-type because some of the applications do not directly relate to strictly dynamic systems. They do, however, relate to solving (1.1) for the input $u(t)$. In [13-15], methods based upon the direct evaluation of inversion formulas for convolution equations with analytically known kernel k are utilised to obviate the difficulties associated with noisy output-type data. However, it is known from [16] that the construction of such inversion formulas requires that either (1.1) can be differentiated n times to obtain a second kind convolution equation or

$$1/[s^n L\{k(t)\}] \text{ is a Laplace transform .} \qquad (3.4)$$

It is shown in [16] that such second kind equations have an inversion formula. Indeed, a large class of convolution equations arise in the field of stereology [13]. Many have kernels $k(t-s)$ which are of the form or can be reduced to the form

$$v(t)w(s)(t-s)^{-\frac{1}{2}}$$

and are called Abel-type equations with separable kernels. Because the known $v(t)$ and $w(s)$ can be absorbed into $x(t)$ and $u(s)$, respectively, they are effectively of simple convolution type. The point we make here is that methods for these equations can be based upon the evaluation of their inversion formulas since

$$1/[sL\{t^{-\frac{1}{2}}\}] = (\pi s)^{-\frac{1}{2}} \text{ is a Laplace transform.}$$

Regularisation and spectral filtering methods which have been used for numerical differentiation [17], for example, represent other approaches used to estimate input-type functions. Indeed, it seems likely that there is some link between the regularisation approach and Kalman filtering mentioned previously, although this has not been investigated. In simple terms, regularisation is the minimisation of the error of a *finite difference* approximation to (1.1) as given, for example, by

$$\begin{aligned} x_i &= k_i u_0 + k_{i-1} u_1 + \dots + k_0 u_i \\ &= K_i(z^{-1}) u_i \end{aligned}$$

subject to a smoothing constraint on the solution. A typical example would be

$$\min \|x_i - K_i(z^{-1})u_i\|_2^2 + \gamma \|u(t)\|_W^2$$

with W being an L_2 or Sobolev norm. The quantity γ is known as the regularisation parameter and is usually of small positive value, often determined by trial and error.

Perhaps a suitable means of circumventing some of the difficulties here is to determine the rational TF model of (1.1), change it to an invertible but dynamically equivalent model, if necessary, and solve the resultant equations subject to a suitable constraint on the solution. This would overcome the problem of the inverse TF method when there is noise on the data but still make use of its parametric efficiency and avoid problems of non-invertibility.

Appropriate choice of the smoothing norm W of the solution is another problem encountered in the implementation of regularisation-type procedures. For example, a popular choice is

$$\|u(t)\|_W^2 = \|u(t)\|_{L_2}^2 \; . \tag{3.5}$$

In Figure 6, we display the solution for such an implementation against what we believe to be a good solution of an impulse response estimation problem discussed in the next section. Intuitively, it can be thought that the effect of a constraint like (3.5) is to reduce the peak values of the true solution and this is apparently the effect. Knight and Selinger [19] also remark on the difficulty of choosing an appropriate smoothing norm for estimating from (1.1) a function describing fluorescence decay. They tried a constraint of $\|u''(t)\|_{L_2}^2$ concluded that 'even for small errors in the data, the method introduced unwanted structure in the solution and with data containing large statistical error, there were no values of γ for which physically sensible results could be obtained'.

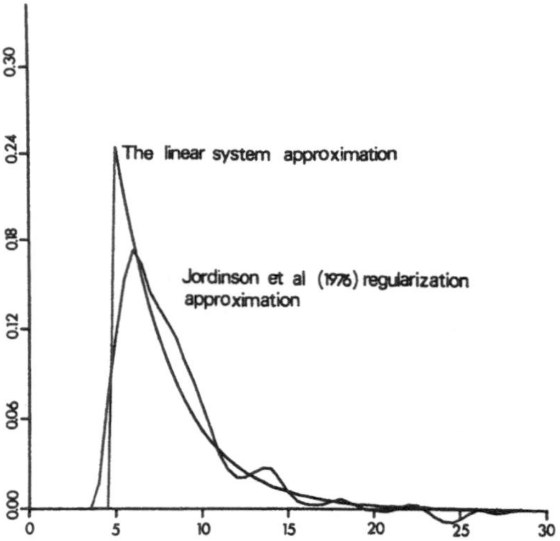

Figure 6 : Comparison of impulse response estimation by the TF based systems approach and a regularisation approach (Jordinson et al [24]) with a smoothing norm of the form (3.5).

4. SOME APPLICATIONS OF IMPULSE RESPONSE ESTIMATION

In this section we briefly mention some interesting applications of the transfer function approach to the impulse response estimation problem and attempt to briefly describe their relevance. All the results relate to discrete-time model estimation although continuous-time results are usually available in the references given.

4.1 Fluorescence decay [20].

In this problem pulse-like inputs of light u_i $(i = 1,\ldots,N)$ are used to excite a compound in liquid or vapour state to produce, over a period of time, an observed probability density y_i $(i = 1,\ldots,N)$ of fluorescence decay with respect to the independent variable wavelength. In this case, it is the true fluorescence decay given by $k(t)$ in equation (1.1) in which we are interested. In particular, the fluorescence lifetime(s) τ_i and the amplitude of the impulse response are required. τ_i is the time constant(s) of the system calculated as in (ii) of section 3 and represents the time required for the true fluorescence decay to decrease to the level $1/e$ times its initial amplitude. It can be used, together with other experimentally determined parameters such as quantum yield, to derive energy properties of the excited state. This information is an aid to understanding the mechanism of excitation and deactivation processes.

A given compound may possess more than one decay (n (for discrete time models) or r (for continuous time models) > 1) from an excited state and it is for this reason that the dynamic order of the system, the value of n or r, be identified correctly. Hence the IV identification stage is most useful in this application and is described fully in [20]. For a given experiment from [20] we have reproduced in Figure 7 the fit to the observed fluorescence output, obtained from the operation of the estimated transfer function on the input.

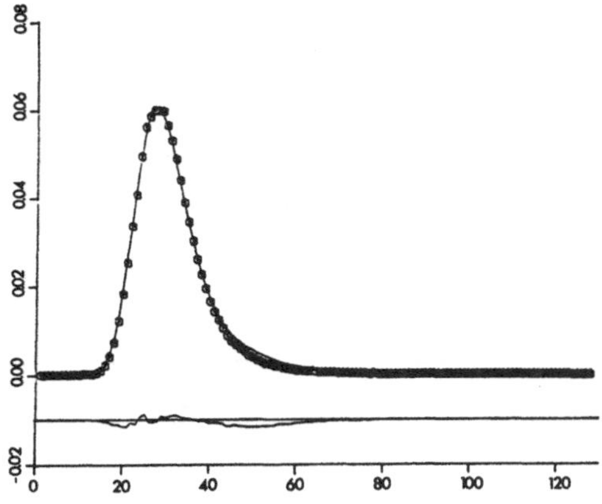

Figure 7 : Comparison of the TF model and actual outputs for the fluorescence of azulene-h8.

4.2 Transport functionals of the circulation in man [21].

This problem can be formulated in the following way: from the injection of indicator or "tracer" into the circulatory system, a pair of indicator curves u_i and y_i are recorded at different sites downstream; the impulse response $k(t)$ in (1.1) is the transport function, between these sites, which is to be calculated. However, the quantitative properties which are of particular interest in this application are related to moments of $k(t)$, that is mean and variance. By realising that the moment generating function of $k(t)$ is its Laplace transform, simple relationships can be obtained between the above properties and the estimated TF parameters [21]. A regularisation solution to this problem was given by [22] and this is plotted in Figure 6 against the impulse response obtained from use of the two stage TF procedure.

4.3 Transport properties of the phloem in plants [23].

This is a similar problem to the above. However, the properties of interest are different. Here, they are mean transport speed and pathway loss. We were supplied with radioactive tracer data from plant experiments conducted by Peter Minchin of DSIR, New Zealand. The results of our IV analysis are illustrated in Figure 8 which shows the actual and predicted tracer concentrations at a point in the plant, based upon the input of tracer material "upstream" of the measurement point. Minchin has since used this approach successfully in a number of experiments on plants [23].

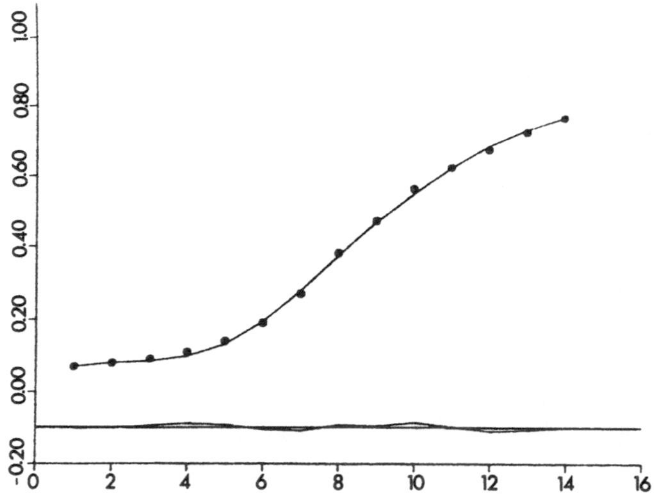

Figure 8 : Forecast of tracer concentration in the phloem of a plant based on a transfer function model between upstream and downstream concentrations.

4.4 Hydrological applications [9,24].

Two commonly occurring examples here are the estimation of rainfall-flow characteristics in catchments and the estimation of dispersion characteristics in rivers. The latter problem is simply the hydrological counterpart of the two biological applications just mentioned. The former problem represents the easier transfer function estimation aspect of the simple running example discussed in sections 2 and 3. In this case, we can obtain an effective rainfall from actual rainfall by taking into account, for example, soil moisture and evaporation effects [9]. It is then a simple matter to obtain the unit hydrograph or impulse response from the transfer function estimated between this effective rainfall and the measured run-off flow. Such a transfer function model is of obvious use for predicting the run-off flow from a catchment for a given rainfall input. A typical example of this is shown in Figure 9.

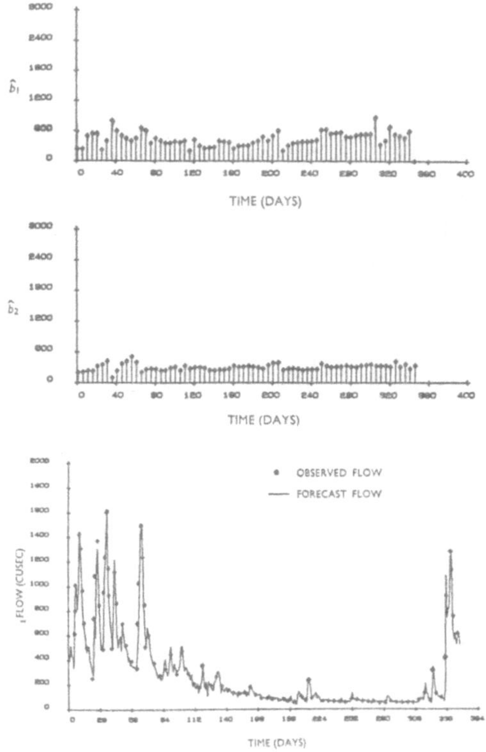

Figure 9 : Forecast of river flow on the Bedford-Ouse River in England based on transfer function model between "effective rainfall" and river flow.

Note that while the relationship between effective rainfall and run-off is linear in this example, the relationship between actual rainfall and run-off is non-linear because of the non-linearity in the effective rainfall computation. This demonstrates that while the basic IV procedure discussed in this paper is appropriate to linear systems, it can be used for non-linear dynamic systems which are linear-in-the-parameters [24].

Amongst many other possible hydrological applications are the determination of transfer characteristics between upstream and downstream flow gauging stations, the estimation of the relation between tide gauge height measures in estuarial locations and the evaluation of transfer functions between input flows and reservoir levels. An example of the upstream-downstream flow application is shown in Figure 10. Here, the downstream flow predicted from a simple transfer function time-series model is compared with the actual downstream measured flow on the Magela Creek in the Northern Territory. The transfer function in this case was of the form

$$\frac{B(z^{-1})}{A(z^{-1})} = \frac{.535z^{-1} + .527z^{-2}}{1-.387z^{-1} + .086z^{-2}} .$$

It can be seen that the steady state gain of this transfer function is approximately $1.5 \left(\frac{.535 + .527}{1-.387 + .086}\right)$ which indicates that about fifty percent more flow is being measured downstream than upstream. The additional flow is clearly due to rainfall-run-off between the gauging stations. This example thus indicates how the simple time-series model can be used to quickly assess hydrological system characteristics.

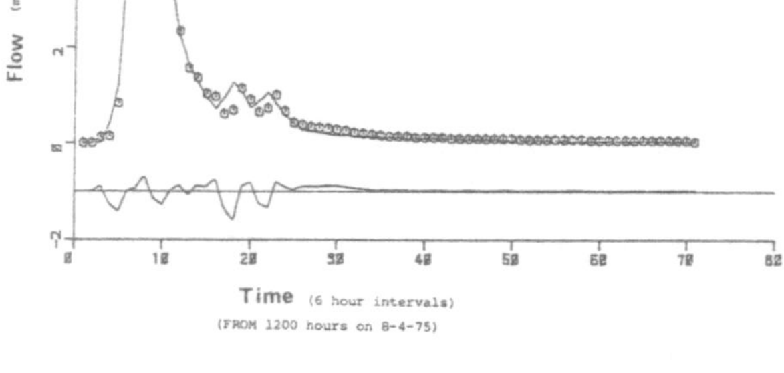

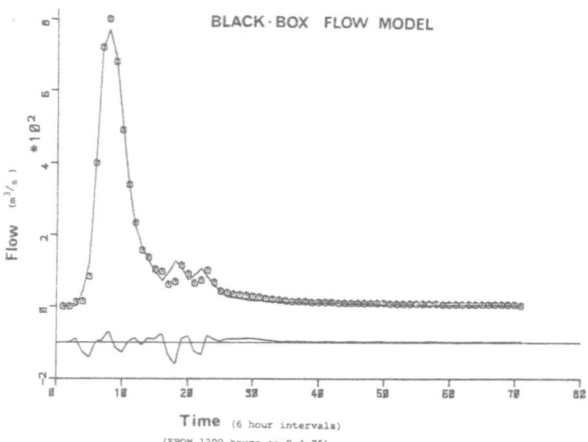

Figure 10 : Forecast of flow on the Magela Creek, N.T., based on transfer function model between upstream and downstream flow.

4.5 Engineering Applications [25].

The whole basis of the transfer function method was evolved for control and systems problems: in such applications it is essential that a parametrically efficient representation is used whenever possible and, for linear systems, the transfer function is the obvious choice. For this reason many engineering applications could be cited and the reader is referred to [25] for some typical ones. Figure 11 shows the results obtained from the analysis of a famous set of gas-furnace data, originally analysed by Box and Jenkins [26].

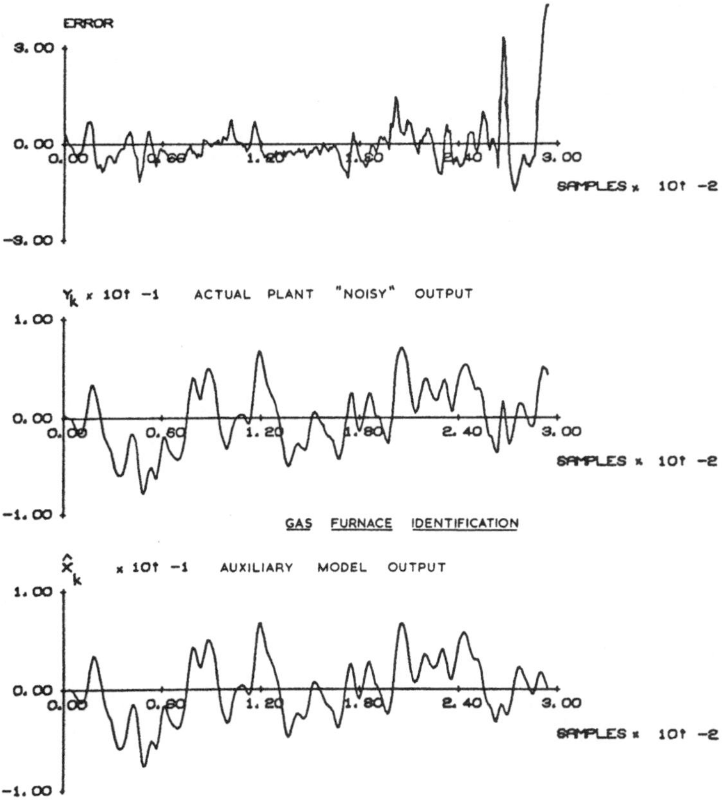

Figure 11 : Comparison of transfer function model and actual outputs for a famous gas-furnace.

4.6 Time-Series and Econometric Applications [27]

Parallel with developments in the control and systems field, statisticians have developed time-series procedures which are very similar to those used by control and systems analysts. Principal amongst the proponents of this approach are Box and Jenkins who caused quite a revolution in the field of time-series analysis with the publication of their book on the subject in 197 [26]. Clearly the transfer function time-series estimation methods of systems analysts or the equivalent time-series methods of statisticians can be applied to almost any sets of time-series which exhibit temporal relationships with each other. As a result, program packages which carry out such analyses have been developed in recent years. The programs of Box and Jenkins, for example, are marketed by ISCOL in the UK and our own CAPTAIN package of recursive IV estimation programs are marketed by the Institute of Hydrology and Oxford Systems Associates in the U.K. Figure 12 gives a typical example of the results obtained from the analysis of economic data [27]: here an estimated transfer function model between gross domestic product (GDP) and unemployment is used to obtain forecasts for unemployment.

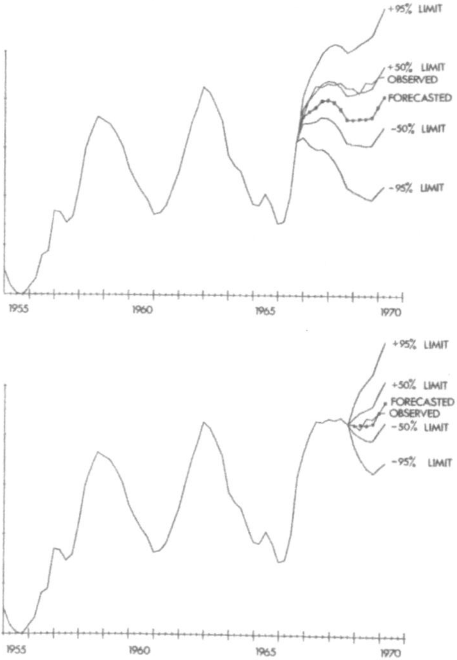

Figure 12 : Forecasts of UK unemployment from (a) 1966 and (b) 1968 based on stochastic transfer function models of GDP-unemployment.

CONCLUSIONS

In this paper, we have seen how a systems approach to time-series analysis can be used to obviate certain problems associated with the solution of convolution integral equations. The approach is directly applicable to the problem of estimating the impulse response or kernel associated with the convolution integral for linear systems and can be used in certain non-linear situations where there is linearity-in-the-parameters. The recursive nature of the IV estimation algorithm is helpful in this latter regard since it allows for the identification of the non-linear phenomena.

A similar approach is of potential use for solving the 'inverse' problem of estimating the system input signal, given an estimate of the kernel and a noisy measure of the output signal. Here the solution is based on the recursive Kalman filter and can be applied directly to problems in which the system model is of a special linear form and the input signal can be described by a Gauss-Markov stochastic model. In such circumstances, the input signal estimate has optimal (minimum variance) properties. In more general situations, the estimation problem is more difficult to solve but a related recursive estimation approach seems possible.

Finally, the paper has provided a number of practical examples from many different areas which demonstrate the efficacy of the systems approach and, in particular, the IV method of recursive estimation.

ACKNOWLEDGEMENT

The authors thank John Childs of the Royal Military College, Duntroon for suggesting the hydrological input estimation problem and for kindly supplying the data used in sections 2 and 3.

APPENDIX

ESTIMATION BY RECURSIVE INSTRUMENTAL VARIABLES

Consider the structural model (2.9) in the scalar form

$$y_k = \frac{b_0 + b_1 z^{-1} + \ldots + b_n z^{-n}}{1 + a_1 z^{-1} + \ldots + a_n z^{-n}} u_k + \xi_k \; . \tag{A1}$$

It is possible to describe (A1) by the following vector relationship

$$y_k = \underline{z}_k^T \underline{a} + \eta_k$$

where

$$\underline{z}_k^T = \left[-y_{k-1}, \ldots, -y_{k-n}, u_k, \ldots, u_{k-m}\right]$$

$$\underline{a} = \left[a_1, \ldots, a_n, b_0, \ldots, b_m\right]^T$$

and

$$\eta_k = \xi_k + a_1 \xi_{k-1} + \ldots + a_n \xi_{k-n} \; .$$

A *least squares approach* to the estimation of $\underset{\sim}{a}$ is to assume η_k uncorrelated with $\underline{z}_k$ and form the appropriate normal equations by minimising $\sum\limits_k \eta_k^2$; viz.

$$\hat{\underline{a}} = \left(\sum_k \underline{z}_k \underline{z}_k^T\right)^{-1} \sum_k \underline{z}_k y_k \; .$$

Suppose that we construct an estimator of the form

$$\hat{\underline{a}} = \left(\sum \hat{\underline{x}}_k \underline{z}_k^T\right)^{-1} \sum \hat{\underline{x}}_k y_k \; . \tag{A2}$$

Since the true value of $\underset{\sim}{a}$ can be given by the expression

$$\underset{\sim}{a} = \left(\sum \hat{\underline{x}}_k \underline{z}_k^T\right)^{-1} \sum \hat{\underline{x}}_k y_k - \underline{x}_k \eta_k \quad ,$$

it follows that

$$\hat{\underline{a}} = \underset{\sim}{a} + \left(\sum \hat{\underline{x}}_k \underline{z}_k^T\right)^{-1} \sum \hat{\underline{x}}_k \eta_k \; .$$

Let us now constrain

$$N^{-1} \sum \hat{\underline{x}}_k \underline{z}_k^T \quad \text{and} \quad N^{-1} \sum \hat{\underline{x}}_k \eta_k \tag{A3}$$

to converge in probability to their expected values, $E_{\hat{x}z}$ and $E_{\hat{x}\eta}$, respectively, and $E_{\hat{x}z}$ to be non-singular. Then

$$\underline{\hat{a}} \xrightarrow{pr} \underset{\sim}{a} + (E_{\hat{x}z})^{-1} E_{\hat{x}\eta} .$$

Therefore, $\underline{\hat{a}}$ is *consistent* when $E_{\hat{x}\eta}$ is zero. When $\{\xi_k\}$ and hence $\{\eta_k\}$ is a correlated sequence, $\{\hat{x}_k\}$ must be independent of $\{\eta_k\}$ so that $\underline{\hat{a}} \xrightarrow{pr} \underset{\sim}{a}$. In the least squares case, $\hat{x}_k$ is given by $\underline{z}_k$ which is not independent of $\{\xi_k\}$ unless the latter sequence is uncorrelated.

However, the estimator (A2), when the terms in (A3) are constrained to satisfy

$$\det E_{\hat{x}z} \neq 0 \quad \text{and} \quad E_{\hat{x}\eta} = 0 , \tag{A4}$$

is consistent and is known as an *instrumental variables estimator*. To obtain such an estimator, the instrumental variable vector $\underline{\hat{x}}_k$ can be chosen to be as highly correlated as possible with the hypothetical noise-free vector $\underline{x}_k^T = \left[-x_{k-1} \ldots -x_{k-n} u_k \ldots u_{k-m}\right]$ satisfying (A4), but totally uncorrelated with the noise $\{\xi_k\}$.

In practice, the vector $\underline{\hat{x}}_k$ is constructed by using an *auxiliary model* of the system. Such an approximation of the system model is obtained by an a priori estimate of $\underset{\sim}{a}$ using, for example, an initial least squares solution. This is updated either recursively or iteratively [5] as better estimates become available. The vector $\underline{\hat{x}}_k$ is given by

$$\underline{\hat{x}}_k = \left[-\hat{x}_{k-1} \ldots -\hat{x}_{k-n} u_k \ldots u_{k-m}\right]$$

where

$$\hat{x}_k = \underline{\hat{x}}_k^T \underline{\hat{a}}$$

is the auxiliary model output of the system using the auxiliary model $\underline{\hat{a}}$.

Finally, it can easily be shown [3] that the recursive form of an instrumental variables solution is

$$\underline{\hat{a}}_k = \underline{\hat{a}}_{k-1} - \underline{w}_k \{\underline{z}_k^T \underline{\hat{a}}_{k-1} - y_k\}$$

$$\hat{P}_k = P_{k-1} - \underline{w}_k \underline{z}_k^T \hat{P}_{k-1}$$

$$\underline{w}_k = \frac{\hat{P}_{k-1} \underline{\hat{x}}_k}{1 + \underline{z}_k^T \hat{P}_{k-1} \underline{\hat{x}}_k}$$

REFERENCES

[1] Takahashi, Y., Rabins, M.J., and Auslander, D.M., *Control and Dynamic Systems* : Addison-Wesley, London, 1972.

[2] Young, P.C., Regression analysis and process parameter estimation - a cautionary message. *Simulation* 10, 3, 125-128, 1968.

[3] Young, P.C., Recursive approaches to time series analysis. *Bull. Inst. Maths. Appl.* 10, 209-224, 1974.

[4] Young, P.C., Some observations on instrumental variable methods of time series analysis. *Int. J. Control* 23, 593-612, 1976.

[5] Young, P.C., and Jakeman, A.J., Refined instrumental variable methods of recursive time series analysis, Part I : single input, single output systems. *Int. J. Control*, 29, 1-30, 1979.

[6] Jakeman, A.J., and Young, P.C., Refined instrumental variable methods of recursive time series analysis, Part II : multivariable systems. *Int. J. Control,* 29 621-644, 1979.

[7] Young, P.C., and Jakeman, A.J., Refined instrumental variable methods of recursive time series analysis, Part III : Extensions. CRES Report Number AS/R17 (Australian National University) 1978. (to appear in *Int. J. Control*).

[8] Young, P.C., Jakeman, A.J., and McMurtrie, R.E., An instrumental variable method for model structure identification. CRES Report Number AS/R22, (Australian National University) 1978. (to appear in *Automatica*).

[9] Whitehead, P.G., and Young, P.C., A dynamic stochastic model for water quality in part of the Bedford-Ouse river system. In : *Modeling and Simulation of Water Resources Systems,* G.C. Vansteenkiste (ed.) North Holland : Amsterdam, 1975.

[10] Kalman, R.E., a New approach to linear filtering and prediction theory. *Trans. Am. Soc. Mech. Engng., J. bas. Engng. Div.* 82, 35-45, 1960.

[11] Young, P.C., and Jakeman, A.J., An inverse problem : the estimation of input variables in stochastic dynamic systems. CRES Report Number AS/R28, 1979.

[12] Kaldor, J.M., The estimation of parametric change in time series models. *M.A. Thesis,* Centre for Resource and Environmental Studies, Australian National University, 1978.

[13] Jakeman, A.J., and Anderssen, R.S., Abel type integral equations in stereology I : general discussion. *J. Microscopy,* 105, 121-133, 1975.

[14] Anderssen, R.S., and Jakeman, A.J., Abel type integral equations in stereology II : computational methods of solution and the random spheres approximation. *J. Microscopy* 105, 135-153, 1975.

[15] Anderssen, R.S., Computing with noisy data with an application to Abel's equation. *Error, Approximation and Accuracy* (ed. by F.R. de Hoog and C.L. Harvis), Queensland University Press, 1973.

[16] Jakeman, A.J., Numerical inversion of Abel type integral equations in stereology, *Ph.D. Thesis*, Computer Centre, Australian National University, 1975.

[17] Anderssen, R.S., and Bloomfield, P., Numerical differentiation procedures for non-exact data. *Num. Math.* 22, 157-182, 1974.

[18] Henrici, P., Discrete Variable Methods in *Ordinary Differential Equations*, John Wiley, New York, 1962.

[19] Knight, A.E.W., and Selinger, B.K., The deconvolution of fluorescence decay curves. A non-method for real data. *Spectrochimica Acta*, 27A, 1223-1234, 1971.

[20] Jakeman, A.J., Steele, L.P., and Young, P.C., A simple transfer function method for estimating decay functions and its application to fluorescence decay data. CRES Report Number AS/R26, Australian National University, 1978.

[21] Jakeman, A.J., Young, P.C., and Anderssen, R.S., Use of the linear systems approach for transfer function analysis of portions of the central circulation in man. CRES Report Number AS/R23, Australian National University, 1978.

[22] Jordinson, R., Arnold, S., Hinde, B., Miller, G.F., Rodger, J.G., and Kitchin, A.H., Steady state transport function analysis of portions of the central circulation in man. *Comput. Biomed. Res.* 9, 291-305, 1976.

[23] Minchin, P., Analysis of tracer profiles in phloem transport : separation of loading and transport phenomena to obtain a mean transport speed and pathway loss (available from author, Physics and Engineering Laboratory, Department of Scientific and Industrial Research, Lower Hutt, New Zealand), 1978.

[24] Young, P.C., A general theory of modeling for badly defined systems, CRES Report Number AS/R9, Australian National University, 1977. (also in *Modeling, Identification and Control in Environmental Systems*, G.C. Vansteenkiste (ed.) North Holland, American Elsevier, 1978, pp.103-136).

[25] Young, P.C., Lectures on recursive approaches to parameter estimation and time series analysis, in : *Theory and Practice of Systems Modeling and Identification*, ONERA - Toulouse Research Centre, Tolouse, 1971.

[26] Box, G.E.P., and Jenkins, G.M., *Time Series Analysis, Forecasting and Control*, Holden Day, San Francisco, 1970.

[27] Young, P.C., Naughton, J.J., Neethling, C.G., and Shellswell, S.H., Macro-economic modeling - a case study. IN : *Identification and System Parameter Estimation*, P. Eykhoff (ed.) New York: American Elsevier, Amsterdam : North Holland, 145, 1973.

INDEX